A Conception of Teaching

T0204647

Nathaniel L. Gage

A Conception of Teaching

 Springer

Dr. Nathaniel L. Gage
Stanford University
School of Education
Stanford, CA
USA

ISBN: 978-0-387-09445-8 (hardcover)   e-ISBN: 978-0-387-09446-5
ISBN: 978-0-387-84931-7 (softcover)
DOI: 10.1007/978-0-387-09446-5

Library of Congress Control Number: 2008931138

Printed on acid-free paper

springer.com

To Maggie

Since March 10, 1940

Professor Nathaniel L. Gage
1917–2008

Tribute

"This is an important book by the top scholar on research on teaching. As always, Professor Gage has much to say–and he says it well."

TOM GOOD
PROFESSOR OF EDUCATION, UNIVERSITY OF ARIZONA

"The scope and import of N. L. Gage's scholarship is easily documented. Others have described this. Less tangible or apparent is his impact on those who had the immeasurable good fortune to enjoy his mentorship. He modeled devotion to the disciplines of scholarship – a never satisfied striving toward thoroughness, dedication to clarity especially about methodologies that color interpretations, insistence on clarifying foggy arguments, and commitment to stimulating rather than disparaging other scholars' thinking. His extraordinarily high expectations blended naturally with gentleness in corrections and patience in explaining not just what was the case but how a case came to be the case. These qualities were all the more special in the context of almost always being much too near a deadline. Like other superb mentors, Nate taught students and colleagues a very great deal about educational psychology. More significantly, he led us to learn and prize the ethic of scholarship."

PHIL WINNE
PROFESSOR & CANADA RESEARCH CHAIR, SIMON FRASER UNIVERSITY

"Besides being a superb scholar and an exacting editor, Nate Gage was a consummate gentleman of the old school. I remember a meeting of the "Invisible College" at Syracuse University in the early 1980s. Nate was reporting on the procedures he had used in his planning and preparation for the National Institute of Education's 1974 Conference on Studies in Teaching, which he chaired. He showed a slide of a chalkboard used during the planning process. On the chalkboard he had listed the ten panels that were to make up the conference, along with the chairs and members of each panel. I was struck by the fact that all the male participants were identified by last name only, while the female participants were identified by both first and last name. Given the concerns about gender equity of that time period, I wondered if this could be an instance of sexism. But when I asked him to explain the difference in recording of the names, he said that he thought it would be rude to refer to women by their last name only. He clearly cared about issues of gender equity. When Ann Lieberman and I organized a special workshop on problems of two-career marriages at a later AERA conference, Nate attended, and confessed to the assembled group that he felt guilty about the ways that he had probably inhibited Maggie, his dear wife, from exploring her own career possibilities. He knew that his early views on the role of women were constrained by the social expectations of that period, and he really worked at understanding and accepting the newer perceptions of appropriate roles. The number of women scholars who served as members of the NIE panels reflected that new understanding. His chalkboard notes demonstrated how his early views of appropriate ways of addressing women were incorporated into his newer views of appropriate roles for women. I have long cherished the memory of that chalkboard image. For me, it captures one of the ways Nate's personal beliefs contributed to his professional relationships."

GRETA MORINE-DERSHIMER
EMERITUS PROFESSOR OF EDUCATION, UNIVERSITY OF VIRGINIA

"I met Nate Gage in August of 1974. I was a newly admitted doctoral student in educational psychology at Stanford University, and Nate was my academic advisor. We sat together in Cubberly Library reviewing my resume, and he asked me how I came to know my previous employer who had written a letter of recommendation for my file. I told him that Bob Baker's daughter was my best friend in high school and after we graduated college, and Bob asked if I would like to come to California and work in his research lab. Nate responded with a chuckle, and said with a twinkle in his eye, "Well, you must have impressed him in some other way because he could not write a letter like this otherwise." Nate still had this same, warmly playful spirit the last time I saw him as well. Almost exactly 33 years later, in August of 2007, Nate appeared unexpectedly, walking down the aisle with his assistant, at my outgoing address as the president of Division 15 (Educational Psychology) of the American Psychological Association. I was so moved that he would make the effort at 90 plus years of age to come to San Francisco to hear me speak that I had to take a moment, pause, and introduce him to the many younger faces in the audience. That same twinkle was there in his eye then, and the last thing he said to me as he left the reception afterward was, "Baby girl, you did me proud." Perhaps the most meaningful thing I can say about Nate Gage is simply that *he was there for me* throughout my career."

LYN CORNO
TEACHERS COLLEGE, COLUMBIA UNIVERSITY

NATE GAGE: A MAN FOR ALL SEASONS

"Although Nate was known as a psychologist, his work encompassed much more – some philosophy of education, some history of education, some sociology of education. All aspects were brought to bear in his attempt to create a theory of teaching that was both rigorous and based on empirical evidence. It was his quest for empirical evidence that brought us together. Nate joined the Stanford faculty in 1962, a year after I arrived. We often discussed some of the research questions in education. He posed a problem that often plagued social scientists, namely, that although there may be numerous studies on a particular subject, the evidence from any single study may not be convincing. This might be due to small sample sizes (often the case in dissertations), or there may be heterogeneity in the data. However, the composite of the studies seem to point in one direction. Nate's question was whether this disparate evidence could be combined in a statistically rigorous way so as to yield a convincing conclusion. Statistical methods for combining the results of independent studies have been called meta-analysis, and Nate's question was a catalyst for me to begin working on the development of statistical methods for the analysis of such data.

Nate had a kind sense of humor. He had a number of sentences that he liked and which he would repeat. One such would come out at faculty meetings after I would make some comment involving numbers. Nate would say, "Ingram, you know that you are no good with numbers." This was also said when we had to check the bill at a restaurant.

For the last 45 years Nate has been a friend, a colleague, and an intellectual stimulus. Most recently, he expressed concern that the education community recognize that teaching had a scientific foundation. This book represents his legacy in showing that such a foundation exists."

INGRAM OLKIN
PROFESSOR OF STATISTICS AND EDUCATION, STANFORD UNIVERSITY

"When I first met Nate Gage in 1972, he was described as the "father of research on teaching." I didn't quite know what that meant at the time, being a new staff member at the National Institute of Education, just completing my dissertation on students' achievement motivation. I soon learned. Nate Gage worked relentlessly on developing and sustaining research on teaching and bringing to it the prestige required to be accepted as an educational research field. Nate's strong focus on research on teaching, its concept, conduct, power, and use continued throughout his lifetime. He was a humble man: one who never placed himself in front of the pack, and would converse with anyone—young or older academics, school people, politicians—with respect and grace. His knowledge of the field was simply remarkable. He was, indeed, the Father of Research on Teaching, and will be sorely missed."

VIRGINIA RICHARDSON
PROFESSOR OF EDUCATION EMERITA, UNIVERSITY OF MICHIGAN

"Nate was a mentor and safe harbor during my graduate student days at Stanford (1968-1970) and then a colleague and good friend throughout the remainder of his career. Many memories come flooding back, but two stand out as a graduate student and are still vivid. The first memory is that of a big fight Nate and I got into at a meeting of the Psychological Studies in Education faculty. He was a chaired professor; I was a graduate student representative. The fight was whether psychological principles were general or subject-matter specific. Nate held the former view and I held the latter. We got into a real match... so much so that Lee Cronbach caught up with me after the meeting saying that it was apparent that senior faculty as well as graduate students could make fools of themselves! The "debate" was quickly forgotten and had no bearing on Nate as a mentor and friend.

The second memory is of Nate at home. Nate and Maggie often invited graduate students over for wine, cheese, and conversation. On one such occasion I learned that Nate was on volume Q of the Encyclopedia Britannica. Nate was reading the encyclopedia from A to Z, no doubt editing as he went along!

Two memories of Nate stand out as a colleague. The first: I had written a paper to present at a conference and asked Nate to read it. He did, editing copiously as was his want. He allowed as how it was a good paper but wasn't ready for publication. I told him I didn't plan to publish the paper. He told me that he never wrote a paper he didn't intend to publish and that stuck with me throughout my career... inhibiting some writing, but, alas, not enough! The second: While I was dean at the Stanford University School of Education (1995-2000), a retired Nate came to talk to me—Nate was working for me, no longer vice versa! He was concerned about finding funding for his next project... a book integrating research on teaching and research on instruction. I suggested he apply for a small grant from the Spencer Foundation. He did so and received funding. He came to my office to report the grant with a smile that made the Cheshire cat look as if it were pouting. And best of all for Nate, his receiving this grant meant he was still in close competition with his brother who, like Nate, was famous in his own field. As fate would have it, our paths crossed in mysterious ways. Nate held the Margaret Jacks chair in Education; I now hold that chair... and it feels good that Nate sat in it as well."

RICHARD J. SHAVELSON
MARGARET JACKS PROFESSOR OF EDUCATION, STANFORD UNIVERSITY

"Good philosophical training cultivates a skeptical eye, and good philosophical training in education often uses educational research to train one's skeptical eye. When I first began to read Nate Gage, I presumed his work would be grist of the mill for my newly cultivated skepticism. Instead, something quite different happened. I realized that his work was "really good stuff," and if I was going to critique it I would have to work very hard. Thus began nearly three decades of back-and-forth exchanges between us that I know were far more beneficial to me than they were for Nate."

GARY D. FENSTERMACHER
PROFESSOR OF EDUCATION EMERITUS, UNIVERSITY OF MICHIGAN

Preface

The only Theory that can be proposed and ever will be proposed that absolutely will remain inviolate for decades, certainly centuries, is a Theory that is not testable. All Theories are wrong. One doesn't ask about Theories, can I show that they are wrong or can I show that they are right, but rather one asks, how much of the empirical realm can it handle and how must it be modified and changed as it matures? (Festinger, 1999, p. 383)

If the history of science proves anything, it is that all Theories prove eventually to be wrong (even if "wrong" only means incomplete or standing in need of elaboration). (Phillips, in Phillips & Burbules, 2000, p. 31)

A word about gender: For millennia, all people, including females, were referred to as "he." I recognize that times have changed by referring to all teachers as "she." Male teachers should accept the error as a compensation for the earlier one.

Acknowledgments

My first three years of work on this book were supported by a small research grant from The Spencer Foundation, which I wholeheartedly thank. For the whole work, the data presented, the statements made, and the views expressed are solely my responsibility. I am, of course, thankful to the scholars and research workers on whose work I have drawn, as noted in the text and list of References. Without the products of their work, especially those of the last half-century, when research on teaching began to flourish, this work would have been impossible.

Professor S. Alan Cohen of the University of San Francisco generously made available to me various publications and other materials that greatly facilitated my becoming informed on the concept of instructional alignment. Several Stanford colleagues answered my queries related to their specialties: Professor Denis C. Phillips, on philosophy of science; Professor Ingram Olkin, on meta-analysis; Professor Decker Walker, on instructional design. The late Professor Kenneth Sirotnik of the University of Washington was altogether cooperative in giving me information about "A Study of Schools," the tour de force led by John Goodlad.

Janet Rutherford, my superb administrative assistant, saved me from months of wandering in the stacks of the Stanford libraries, solved problems with my computer, and provided much highly intelligent general helpfulness. Barbara Celone and Kelly Roll, of Stanford's Cubberley Education Library, and Mary-Louise Munill, of the Interlibrary-Borrowing Services of the Stanford Libraries, obtained many books and articles and, in doing so, occasionally performed seeming miracles.

Dean Richard Shavelson and, later, Dean Deborah Stipek, of Stanford's School of Education, helped me by acting on their faith that, although I was retired, I was making appropriate use of a Stanford University office.

Professors David C. Berliner of Arizona State University, Barak Rosenshine of the University of Illinois, and Raymond L. Debus of the University of Sydney, have been friends of the kind that every author needs. I am immensely grateful for their long-term friendship, encouragement, and criticism. The editorial work of Cynthia Haven saved me from awkwardnesses and obscurities.

Finally, the love and support of my late wife, Margaret, and our children – Elizabeth, Tom, Sarah, and Anne – have kept me steady over the years.

Contents

Chapter 1
An Agenda

After everything else has been done and provided – the money raised; the schools erected; the curricula developed; the administrators, supervisors, and teachers trained; the parents and other citizens consulted – we come to teaching, where all of it makes contact with students, and the teacher influences students' knowledge, understanding, appreciations, and attitudes in what we hope will be desirable ways. Teaching is well-nigh the point of the whole educational enterprise and establishment aimed at producing student learning.

Teaching is also important in terms of a kind of ethical imperative. Nations require that their young people have frequent contact, for long periods, with adults called teachers. When such a relationship is legally imposed on young people, it seems only fair that society should do whatever it can to make that relationship a beneficial one.

The literature of the behavioral and social sciences is full of conceptions and research on learning and memory. Teaching is comparatively a stepchild, neglected by those who have built a formidable body of conceptions of learning and memory. The uses of learning conceptions for teaching constitute a tool-kit that has been left to rust. It is as if the theoretical work of, say, Faraday, had never given birth to the tremendous applications of electrical energy so that when Einstein turned on his lamp, he could read his notes. This book seeks to give teaching the kind of attention that learning and memory have received. Teaching is where learning and memory conceptions should pay off.

Finally, teaching is worth studying simply because of the intrinsic interest of the phenomena to which teaching gives rise. Even if such research had no practical value, it would be worthwhile for the same reasons that astronomy and archaeology are worthwhile. As part of our universe and our human condition, teaching cries out to be studied and understood.

Conceptions are both the guide and the outcome of research, including research on teaching. Research is the process of seeking relationships between variables. That simple definition applies to any science, whether it is in the natural or the behavioral sciences. To explain, we search for logical relationships; e.g., if time is indispensable for learning, lack of time prevents learning. To predict, we search for temporal relationships; e.g., knowing a teacher's high school grade-point average, we can predict with better than chance accuracy, her grade-point average as a college freshman. To control or improve, we search for causal relationships; e.g., knowing that teachers who receive training in question-asking do better than similar teachers

N.L. Gage, *A Conception of Teaching*,
DOI: 10.1007/978-0-387-09446-5_1, © Springer Science+Business Media, LLC 2009

who do not receive such training, we can use that knowledge to bring about better teaching. Explanation, prediction, and control, or one or more of these, are the purposes of all scientific research, including research on teaching.

And what is teaching? We can define it as one person's influence aimed at improving the learning of other persons. Usually, we think of teaching as occurring in face-to-face interaction between the teacher and the learner, but it can also occur when a teacher creates influential events, in which he or she does not participate. In that way, the authors of books and the developers of computer programs may also be considered teachers. But we will restrict our concern to teaching that occurs while a teacher is in the presence of students.

So research on teaching may be defined as the search for relationships between variables where at least one of the variables is a behavior, a thought, or a characteristic of teachers. The teacher variable may be an independent variable, e.g., a way of teaching; or a dependent variable, e.g., the teacher's response to advice; or an intervening variable; e.g., a teacher's thoughts during a student's response to a question, a classroom situation, or some other kind of variable. But at least one teacher variable must be involved if the research is to be research on teaching.

The study of teaching as a concern of the behavioral and social sciences has matured from its philosophical beginnings in antiquity to its present robust youth at the recent turn of the millennium. It is still young, having begun to thrive only during the 1950s. But it is now flourishing with an abundance of scholarly publications by a large number of active researchers on teaching. The result over the centuries, especially the last half-century, has been an accumulation of ideas, concepts, distinctions, insights, empirical findings, and conceptual formulations that seem ready for an attempt at a theory of teaching. Notice the indefinite article: "A." It signifies that mine is just one of an indefinite number of conceivable theories. The various models of teaching described by Joyce, Weil, and Calhoun (2000) could be theorized, i.e., explained in terms of "covering laws," of the kind described in Chap. 8. Why do the authors of the models consider them to be effective, in what ways?

This chapter sketches the development and scope of the conceptions of teaching to be presented. After discussing the choices I have made among various possible emphases and directions, I will summarize each chapter to provide a brief introduction to the rest of the book.

Choices Among Alternative

The theory to be proposed in this book reflects choices made in the early stages of its development.

A Theory of Teaching Rather than Instruction

The differences between the terms *teaching* and *instruction* reside mostly in their connotative meanings. But those differences are clear enough to be relevant to the scope of this monograph. "Teaching" is the term used more in formal educational settings, namely, in elementary schools, secondary schools, colleges, and graduate schools. "Instruction" is used more in sharply focused out-of-school training in business, industry, and the armed forces.

One way to distinguish the two terms was offered by R. M. Gagné and Briggs (1979), leaders in the field known as *instructional design*:

Why do we speak of "instruction," rather than "teaching"? It is because we wish to describe *all* of the events which may have a direct effect on the learning of a human being, not just those set in motion by an individual who is a teacher. Instruction may include events that are generated by a page of print, by a picture, by a television program, or by a combination of physical objects, among other things. Or... the learners may be able to manage instructional events themselves. Teaching, then, may be considered as only one form of instruction, albeit a signally important one (p. 3).

But research on teaching puts teaching, rather than the more general "instruction," at the center because it is the teacher who arranges for the students' interaction with all the media mentioned by Gagné, et al. Typically, the teacher oversees the students in their reading, interaction with computer programs, viewing of films and television, as well as the recitations, discussions, lectures, explanations, and tutoring that occur in schools.

Typically, the teacher directs all aspects of teaching, except for the content of the curriculum, which is usually prescribed for the teacher in varying degrees. The manner, style, and mode of teaching typically fall under the almost complete control of the teacher, especially the use of teaching materials other than the textbook, such as slides, audio tapes, movies, videotapes, digital video displays (DVDs), and computers. Teachers also control the use and arrangement of out-of-school learning experiences, such as excursions and visits to museums.

Instructors have less autonomy; they are more likely to follow the curriculum and materials approved by the organization that employs them. Teachers are formally trained in teacher education programs in colleges or graduate schools. Instructors are usually trained in the business, industrial, or military organization in which they will do their work.

In all of these ways, teaching differs from instruction, not in any formal, legalized, tightly regulated way, but rather in the connotative meanings of the terms as they have come to be used in the United States since at least the mid-nineteenth century when public schools became prevalent.

A Theory of Teaching That is Both Descriptive and Prescriptive

The theory will serve both the descriptive and prescriptive aspects of theory. That is, it will *describe* how teaching does occur and also *prescribe* how it should occur to optimize student achievement.

The idea that there are two kinds of theory, descriptive and prescriptive, is widely accepted (see, for example, Bruner, 1966; Reigeluth, 1999, p. 2). Descriptive theory describes a process as it *does* go on. Prescriptive theory describes how the process *should* go on if it is to be optimized according to some values.

But the distinction blurs when we realize that the same descriptive theory – its concepts and their relationships – can serve both descriptive and prescriptive purposes. That is, when we find that the relationship of variable x to variable y affects an outcome z (a descriptive theory), that relationship can be used to optimize z (a prescriptive theory). The optimization requires that we seek certain values of x and y.

For example, we can describe how teachers explain the Pythagorean Theorem. But if we evaluate the effectiveness of the explanation in terms of student understanding, we can use the explanation-understanding relationship to prescribe how the explanation should be made.

A Conception of Teaching for Both Cognitive and Affective Objectives of Education

The theory of teaching will focus on both the cognitive and the affective objectives of education. Of course, a good deal of teaching, especially in elementary schools, is concerned with the emotional and social development of students as well as with their cognitive development (see R. B. Smith, 1987). Still, teachers' concern with emotional development typically may tend to decrease gradually from the 1st to the 12th grades.

A Broadly Valid, Rather than Specifically Valid, Theory

The theory will apply to many varieties of teaching and have broad validity. It will formulate a set of widely valid concepts or variables to describe teaching and the widely occurring relationships between those concepts or variables. The breadth of the theory signifies its attempt to describe and explain teaching's many dimensions: the teaching of many kinds of *subject matter* at many levels of *student maturity*, toward many sets of *cognitive educational objectives*, to students of any *gender, social class, or ethnicity*, in many *school or classroom settings*, by *many kinds of teachers*, in many *cultures*. B. O. Smith (1963) expressed an even bolder aspiration toward a *universally* valid theory of teaching:

Our most general notion is that teaching is everywhere the same, that it is a natural social phenomenon and is fundamentally the same from one culture to another and from one time to another in the same culture. Teaching is a system of action involving an agent, a situation, and an end-in-view, and two sets of factors in the situation: one set over which the agent has no control (for example, size of classroom and physical characteristics of pupils) and one set which the agent can modify with respect to the end-in-view (for example, assignments and ways of asking questions) (p. 4).

Any attempt at universality in a conception of teaching runs into the great variety of subject matters taught. For example, a book on subject-specific teaching (Brophy, 2001) included 14 chapters, each by a specialist on teaching methods and activities for a specific subject: beginning reading, content area reading and literature, writing, mathematics of number, school geometry, biological literacy, physics, representations, earth science, history, physical geography, cultural geography, citizenship, and economics. Within each of these subjects, there are presumably optimal instructional methods specific to particular kinds of content. We can make further breakdowns for specific kinds of students in terms of their cultural backgrounds, levels of cognitive capability, cultures, communities, and so on.

Against the assumption underlying the Brophy-edited volume is the view of R. M. Gagné (1976): "Learning is not unique to subject matter. There is no sound rational basis for such entities as 'mathematics learning,' 'science learning,' 'language learning,' or 'history learning,' except as divisions of time" (p. 30).

Gage (1979) proposed that the generality-specificity issue be resolved by creating a hierarchy of levels of generality shown in Table 1.1.

The theory to be proposed takes the highly general tack. Although much of this book will seem to have been aimed at only elementary and secondary school teaching, it may also apply to college teaching, as was implied when Bellack (1976) noted that his formulation of the process of teaching had also been observed at the college level (see pp. 5–31). As Sirotnik (1983) observed, we can never understand teaching if we need a separate theory to explain each of the myriad forms that teaching can take in types of subject matters taught, of students, of community contexts, and of resources available.

Table 1.1 Possible levels of generality-specificity for a theory of teaching

Level I	All grade levels, subject matters, student types.
Level II	Major grade-level categories, such as preschool, early primary grades, late elementary grades, secondary, and college levels.
Level III	Major subject-matter categories, such as verbal, mathematical and scientific, aesthetic, and psychomotor.
Level IV	Major grade-level subject-matter combinations, such as primary-grade reading, upper-elementary social studies, high school geometry, and college physics.
Level V	Major grade-level subject-matter combinations for students at different points on various dimensions, such as general cognitive capability, academic motivation, ethnic identity, socioeconomic status, sensory and motor abilities.
Level VI	Major topics within grade-level subject-matter combinations such as the sound of "th," the Bill of Rights, the Pythagorean theorem, Ohm's law.

A Theory of Teaching Actions and Teacher Characteristics

The term "teacher actions" refers to what teachers *do*: explain, ask questions, etc.

The term "teacher characteristics" refers to what teachers *are*: recently trained, experienced, etc.

The term "*teaching* effectiveness" implies that the teacher's *actions*, such as her ways of explaining and questioning, account for her effects on students. The term "*teacher* effectiveness" implies that it is her *characteristics and personality traits,* such as her intelligence, knowledge, and emotional stability, that account for the teacher's effects on student achievement.

Much early research, reviewed by Getzels and Jackson (1963), showed that *teacher-characteristic* variables account for little of the variance in student achievement. More recent studies, however, reverse that trend. For example, Ehrenberg and Brewer (1995) found that teachers' verbal aptitudes boosted student achievement gains. Monk (1994) found that secondary school teachers' preparation in mathematics and science raised student gains in mathematics and science. Strauss and Sawyer (1986) showed that the cognitive abilities of teachers not only affected student achievement but also lowered student dropout rates.

The proposed theory will assume that the teachers' verbal aptitudes, preparation in the subject matter, and cognitive abilities affect their decisions and behavior – all of which influence student achievement substantially. Accordingly, the theory will address both actions in "teaching" and characteristics of the "teacher."

A Theory of Classroom Teaching Rather Than Any of the Challenges to Classroom Teaching

Classroom teaching has long been decried and challenged. John Dewey's progressive education was an early challenger. As Cuban (1992) showed, it never took hold.

B. F. Skinner's (1968) programmed – and, later, computer-assisted – instruction has been greatly strengthened by the ongoing computer revolution, but it still is used by only a minority of teachers (H. Becker, 2000). Cuban (1986, 2001) has continued to find relatively little use of computers in classrooms. In a possible exception to this trend, H. Becker reported *relatively* abundant use of computers by the students of teachers who had at least five computers in the classroom, had some competence with computers, and were well above average in the strength of their belief in a constructivist (described in Chap. 5) teaching philosophy. So only under special conditions of teacher preparation do classrooms appear to be different than they have been for many decades.

Fred Keller's (1968) "personalized system of instruction" in the 1970s, Benjamin Bloom's (1968) "mastery approach" in the 1970s, Ann Brown's (1989, 1996) and others' "reciprocal teaching" in the 1980s, and Robert Slavin's (1990) "cooperative learning" in the 1990s, have not, in the absence of evidence to the contrary, been adopted by the vast majority of 3.5 million U.S. elementary and secondary school teachers (National Center for Educational Statistics, 2000), to say nothing of teachers abroad.

The ubiquity and tenacious survival of conventional-direct-recitation (CDR) teaching (described in Chap. 5) is the main reason for my decision to focus on it to develop a conception. (For a brief summary of the early research on the survival power of conventional-direct-recitation teaching, see Sirotnik 1983.)

Observational studies in the United States (for example, Bellack, Kliebard, Hyman, & Smith, 1966; Hoetker & Ahlbrand, 1969; Mehan, 1979; Goodlad, 1984) have agreed in showing that, when they examine the observational evidence about what goes on in classrooms, conventional-direct-recitation (CDR) teaching prevails. In the *recitation cycle,* (a) the teacher "structures" the subject of the discussion, (b) then asks a question, (c) then either calls on a volunteer or selects a student to respond, and (d) finally reacts to the student's response. Although these cycles are repeated for much of a class period, the teacher may provide time for individual students to work alone or in a small group on an assigned task.

An Overview of Chapters 2–9

The following sections briefly describe the subsequent chapters. They orient the reader to the work as a whole and sketch the context into which I place the elements of the argument.

Chapter 2

The Need for a Theory of Teaching

I first describe the need in terms of the many affirmations from philosophers and behavioral scientists over the years. In the process, I consider whether scientific research requires a *prior* theory – a theory spelled out before any data are collected. The issue is resolved with ideas from *Conjectures and Refutations* by the philosopher of science Karl Popper (1963).

Chapter 3

The Possibility of a Theory of Teaching

This section – an updating of Gage (1996) – is relatively technical and not indispensable to a comprehension of the book as a whole. It deals with the negative responses to the possibility-of-theory question. Among others, two behavioral scientists

– Cronbach (1986), an educational psychologist, Gergen (1973), a social psychologist, and Thomas (2007), an educational psychologist – have argued that, because the requisite raw material of any theory consists of generalizations, and because lasting and broadly valid generalizations in the behavioral sciences are impossible, valid theories in the behavioral sciences or, at least, those in educational and social psychology, are impossible. The chapter rebuts their arguments with logical analysis and empirical evidence. It ends with an examination of their implied indeterminism and a defense of the possibility and value of *probabilistic* theory in the behavioral sciences.

Chapter 4

The Evolution of a Paradigm for the Study of Teaching

The proposed paradigm – or model of a scientific field – comprises six basic categories of related concepts that underlie the proposed theory. Because these concepts can take forms that vary qualitatively, quantitatively, or both, they will also be called *variables*. I describe and illustrate the variables in terms of how they have entered into analytical (logical) and empirical studies of teaching.

The chapter presents these concepts in the historical order in which the categories were developed. But they are arranged spatially in a "pedagogical" order that makes sense for all amounts of teaching – whether they last a few minutes or a school term. In that order, some categories of concepts, logically at least, must precede others

The categories of concepts can be divided into two sets: (a) those that are logically prior to a teacher's teaching, i.e., her presentation of the process-and-content of teaching, and (b) those that logically occur *after* her teaching.

Chapter 5

A Conception of the Process of Teaching

This chapter discusses the thoughts and behaviors of teachers as they seek to foster their students' achievement of the objectives of the teaching. Many models of teaching have attracted some attention, but only one seems to have won the allegiance of the vast majority of teachers in elementary and secondary schools. The chapter describes that model and the evidence of its wide usage – not only recently, but in the whole twentieth century; not only in the United States, but in other countries; and not only in a few subjects, but in many. The chapter also examines the reasons for the persistence and prevalence of this model.

Chapter 6

A Conception of the Content of Teaching

This chapter presents a concept – *instructional alignment* – that has proven useful in a variety of contexts in which the content of teaching has played a part. These contexts consist largely of the roles content plays in teaching and also in assessing achievement. This discussion leads to an examination of the issue of teaching-to-the-test content, as in Popham's (1993) "measurement-driven-instruction," versus testing-to-the-content-taught. The issue is resolved in terms of making the curriculum dominate both teaching and testing.

Chapter 7

A Conception of Students' Cognitive Capabilities and Motivation

Cognitive capabilities consist of (a) the general cognitive ability (IQ) of students, (b) their multiple intelligences, and (c) their prior knowledge of the content being taught. Students in any class differ in their cognitive capabilities. The differences among students raise the problems of maximizing teaching's appropriateness to those differences. The chapter examines ways in which teachers can achieve that appropriateness. At its end, the chapter examines the behavioral and cognitive ways of influencing student motivation.

Chapter 8

A Conception of Classroom Management

The chapter begins with a presentation of the role of poverty in creating much of the problem of classroom management. The problem is much greater in schools and classrooms that serve students from impoverished families and neighborhoods. Classroom management is aimed at the proper use of classroom time: maximizing instructional time and minimizing counterproductive time. The model of classroom learning developed by Carroll (1963) lends itself to focusing on time as a basis for defining student aptitude, opportunity to learn, and perseverance in terms of time.

Students' thought processes are important aspects of instructional time because they provide a criterion of effectiveness in use of time. They are regarded as intervening variables, occurring between teaching and achievement.

Productive and counterproductive use of time can be related to classroom management practices. Research has identified such practices in elementary and secondary schools.

Avoiding various kinds of teacher bias on grounds of student gender, cognitive capability, ethnicity, and socioeconomic status is an important aspect of classroom management.

Chapter 9

Integrating the Conceptions

The chapter begins with the ideas of the philosopher of science Hempel (1965) concerning how a conception can consist of sub-conceptions. I illustrate sub-conceptions with an analogy: four sub-conceptions serving as an explanation of an automobile's motion.

Sub-Conceptions of Process

To optimize process, we must optimize the four components of the classroom-direct-recitation (CDR) model of teaching described in Chap. 5: structuring, soliciting, responding, and reacting. The empirically identified effectiveness of certain forms of these components is in each case regarded as a sub-conception consisting of: (a) a phenomenon in such teaching that calls for explanation and (b) a covering law providing that explanation.

Each of these phenomena is the concern of a sub-conception. Each sub-conception consists of a sample of research on that component that provides "an example of a phenomenon to be explained." Then an explanation of that phenomenon is provided by a "covering law." The conjoining of a phenomenon to be explained and an explanatory covering law occurs repeatedly to provide a sub-conception for each of the four components of the process of teaching.

Sub-Conceptions of Content

Then the chapter presents in a similar way a set of sub-conceptions of the *content* of teaching. The next section of the chapter consists of an integration of the sub-conceptions of process and content. Finally, the chapter introduces cognitive capability and motivation as modifiers of the relationships between process and content.

Chapter 2
The Desirability and Possibility of a Theory of Teaching

Gary Thomas (1997, 2007), a professor of educational psychology at the University of Birmingham. His paper, "What's the Use of Theory?" (1997) and his *Education and Theory: Strangers in Paradigms* (2007) have attracted attention in large part because of the novelty of their message: the undesirability of theory in education. His position is well represented by the following quotation from his book:

> I contend that the allure of theory – and the desire of educators to call their ideas "theory"– rests historically on its success in other fields, most notably natural science. It was from this success that theory drew its epistemological legitimacy. Many educators appeared to have at the back of their minds the idea that theory represented the clearest distillation of intellectual endeavor: the conceptual and methodological cream of the various disciplines from which it had been borrowed. But my argument is that these successes provide no good reason for contemporary education's romance with theory. The domains in which theory has been useful find no congruence in education. Indeed those domains where theory is valuable are more limited than one might imagine, and I plead for more of the methodological anarchy which Feyerabend (1993), the iconoclastic philosopher of science, pleaded for scientific research. (p. 20)

Thomas's position is based on his apparent conception of education as a homogeneous, unstructured entity, similar to a snowball that contains one kind of molecule (H_2O) and no internal structure. Education, on the other hand, has a large number of discrete components, each of which uses its own concepts and purposes. Examples of these components are educational finance, educational architecture, educational personnel, educational curriculum, and educational assessment. AERA (the American Educational Research Association) has a dozen divisions, such as (a) Administration, (b) Curriculum Studies, and, perhaps of greatest interest to readers of this book, (k) Teaching and Teacher Education. And, because members' interests were often more specific than the Division structure allowed, AERA has about 170 Special Interest Groups, such as Accelerated Schools, Brain and Education, Classroom Management, Design and Technology,... Organizational Theory, Teacher as Researcher, and Writing and Literacies. In short, education has an internal diversification of components infinitely more complex than that of Thomas's snowball-like treatment of education. For that reason alone, Thomas's rejection of the relevance of theory to education should not disallow attempts to formulate a theory of teaching.

N.L. Gage, *A Conception of Teaching*,
DOI: 10.1007/978-0-387-09446-5_2, © Springer Science+Business Media, LLC 2009

An acceptable part of Thomas's argument is that theory rests on generalizations. "[G]eneralization – the drawing of an essence – is what theory is about in one of its most important meanings" (p. 57). But then Thomas cites approvingly the statement by MacIntyre (1985) that conventional social science seeks generalizations, or at least "probabilistic" generalizations, and in this, he says, it has failed: "it has provided nothing in the way of generalizable knowledge in two hundred years" (MacIntyre, 1985, p. 57).

Here MacIntyre and Thomas overlook abundant empirical evidence assembled by Gage (1996) and by Lipsey and Wilson (1995), among others. This evidence of stable, consistent generalizations, some of which is presented in Chap. 3 (pp. 3–15 to 3–20), exists in the form of main effects, correlational and experimental, in the behavioral sciences. Where such main effects are complicated by interaction effects that are long-lasting and dependable across many contexts, those interaction effects are welcome as generalizations produced by the behavioral sciences. But if the interaction effects, of whatever order, are not themselves dependable and thus admissible as part of the orderliness sought by science, they cannot be part of science. If the search for dependable interaction effects can never succeed, then we should despair of making science about interaction effects. But if it does succeed, then interaction effects become part of the reason for rejecting despair for the behavioral sciences. Thus the present argument is not anti-the-importance-of-interaction-effects but rather anti-disorder. The evidence cited below indicates that interaction effects are not so strong or ubiquitous as to vitiate the story told by the main effects in at least some of the ecologies often studied by behavioral scientists.

For example, correlational main effects are abundant. Research on individual differences has established important correlations that have been overwhelmingly positive in the thousands of instances in many contexts since such correlations were first determined early in the twentieth century. Thus, it is hard to find a zero or a negative correlation between two different tests of general mental ability; between the verbal, mathematical, and spatial special abilities; between general mental ability and educational achievement (measured with grades or test scores), between socio-economic status and academic ability; between years of schooling and amount of knowledge; between socioeconomic status and educational achievement.

The status-achievement correlation should remind us that a generalization may change without impairing the scientific status of the underlying knowledge. A case in point: much contemporary educational research seeks to lower that correlation to zero, so that low-resource-family children will do as well in school as middle-class children. But, if that research succeeds, it will not invalidate the underlying generalizations that connect social class to achievement – via the mediating varia-bles in home and school environments, parent-child interactions, and teaching practices (Berliner, 2005). Changes in the ecological frequency or magnitude of certain conditions may change the mean values of these mediating variables with-out changing the mediating relationships involved. (This point was recalled for me by Richard C. Anderson, personal communication, October, 26, 1994, and by Michael J. Dunkin, personal communication, November 10, 1994).

Experimental Main Effects

In experiments, one or more of the independent variables is manipulated to determine whether the manipulation affects other (dependent) variables. Such main effects are also widely discernible. Suppose the generalizability criterion is defined as at least 90% consistency in direction of a relationship over populations of studies numbering at least 20. Reviewing replications of experiments only in research on teaching, Walberg (1986) found such generalizability in the direction of the main effects of (a) cooperative learning on race relations, (b) mid-semester rating-feedback to teachers on improvement in the final rating of the teacher, (c) the personalized system of instruction on student achievement, (d) academic engaged time on learning, (e) innovative curricula as against traditional curricula on innovative learning, (f) behavioral instruction as against traditional instruction on learning, (g) open education as against traditional teaching on attitude toward school, (h) adjunct questions on recall, and (i) student-centered as against instructor-centered discussion on students' attitudes.

A Major Review of Experimental Studies

Lipsey and Wilson (1993) brought together 302 meta-analyses of experimental or quasi-experimental studies of the efficacy of mental health treatments, work-setting interventions, and educational methods. They found that the meta-analytic reviews show a strong, dramatic pattern of overall positive effects that cannot be readily explained as artifacts of meta-analytic technique or generalized placebo effects; indeed such [artifactual or placebo] effects are rather modest" (1993, p. 1181).

They then refined their compilation by eliminating possibly biased treatment-effect estimates – those based on one-group pre-and-post designs, on only published rather than both published and unpublished studies, and on less than the largest number of studies in a given domain. The mean effect size of the remaining 156 meta-analyses was 0.47 (median = 0.44); 99.4% of the effects were positive, and 83% of the effect sizes were 0.20 or higher. More important, such results were found within each of the three categories of interventions. Such effect sizes betoken a highly consistent-in-direction and substantial mean-level of effectiveness. They suggest that, whatever the moderating or weakening influence of interaction effects may be, many main effects are consistent and strong enough to allay despair for the behavioral sciences.

The criticisms of the Lipsey-Wilson work by Sohn (1995) and Eysenck (1995) were countered by Lipsey and Wilson (1995) so as to leave unaffected the present point concerning the magnitude and consistency of main effects. Even if, as Eysenck argued, the psychotherapy effects were largely suggestion effects, the results would support strong and consistent generalizations.

Evaluations of consistency are facilitated by comparisons. Thus, how does the consistency of research results in the behavioral sciences compare with that in the physical sciences? Hedges (1987) made such a comparison by applying an index (Birge's R) of the variability of findings in six behavioral-science research areas: sex differences in (1) spatial and (2) verbal ability, the effects of (3) open education and (4) desegregation on achievement, (5) the validity of student ratings of college faculty, and (6) the effects of teacher expectancy on IQ.

He also determined such consistency of results in 13 areas of particle physics: the lifetimes of nine subatomic particles and the masses of four of those particles. Hedges (1987) found that

> Neither criterion [average value or statistical significance of the consistency measures] indicates a very large difference between the consistency of research results from the social sciences and the consistency of those from the physical sciences. (p. 449)

He also showed that the particle-physics consistencies were similar to other physical science consistencies (e.g., in atomic weights, constants in quantum mechanics) and that the behavioral science consistencies were similar to other behavioral science consistencies (e.g., in gender differences, effectiveness of methods of teaching writing, the validity of personnel selection tests).

On consistency of results, then, the widespread negative view of the behavioral sciences is questionable. The behavioral-science main effects that Hedges examined were about as consistent as the results found in one branch of physics. Even if particle physics is weaker than most physical sciences in its consistencies, the fact that any behavioral science consistencies can come close to those of any respected physical science should alleviate despair.

Evaluating the Magnitude of Main Effects

The magnitude of effect sizes (correlation coefficients or standardized mean differences) in the behavioral sciences has often been derogated. Surely, it is assumed, correlations of 0.15 or effect sizes of 0.20 have no value in theoretical or practical work. But as Rosenthal (1994) noted, "neither experienced behavioral researchers nor experienced statisticians had a good intuitive feel for the practical meaning of common effect size estimators" (p. 242). One basis for such a feel can be obtained by examining main effects obtained in medical research on matters important for theory and practice. The medical main effects were found through correlational studies and randomized experiments based on the same logic as the behavioral science studies that yielded the results mentioned above. Thus, as Table 2.1 shows, the percentage of 518 patients with lung cancer in a hospital who were also smokers was 96%, whereas the percentage of 518 matched patients with other diseases who were smokers was 89% (Lilienthal, Pedersen, & Dowd, 1967, p. 78). The difference of 7% was considered important.

An experiment (Beta-Blocker Heart Attack Trial Research Group, 1982) determined the effectiveness of a drug (propranolol) in reducing fatalities from heart attacks in men who had already had a heart attack. As shown in Table 2.2, of the

Table 2.1 Frequency of smoking in lung cancer patients and matched controls

Smoking status	Patients with lung cancer ($N = 518$)	Matched patients with other diseases ($N = 518$)
Smokers	96%	89%
Nonsmokers	4%	11%
Total	100%	100%

From *Cancer Epidemiology: Methods of Study* (p. 78), by Lilienthal et al. (1967). Adapted with permission.

Table 2.2 Results of beta-blocker trial

After 30-month follow-up	Propranolol ($N = 1,900$)	Placebo ($N = 1,900$)
Dead	0 7.0%	0 9.5%
Alive	0 93.0%	0 90.5%
Total	0 100.0%	0 100.0%

Adapted from "A Randomized Trial of Propranolol in Patients with Acute Infarction. I. Mortality Results" by Beta-Blocker Heart Attack Trial Research Group (1982) Copyright 1982 by American Medical Association. Adapted with permisson.

Table 2.3 Effects of cholesterol lowering

Results after 9 years	Cholesterol lowering treatment ($N = 1,906$)	Placebo
Definite fatal or nonfatal heart attack	8.1%	9.8%
No definite fatal or nonfatal heart attack	9 1.9%	90.2%
Total	100.0%	100.0%

Adapted from Lipid Research Clinics Program (1984).

1,900 men randomly assigned to the group that received propranolol, 7% had died after 30 months, while of the 1,900 in the group that received a placebo, 9.5% had died. The difference of 2.5% between the two percentages was taken very seriously as support for underlying theory and as a basis for medical practice.

In another experiment (Lipid Research Clinics Program, 1984), 3,806 men were assigned at random to a cholesterol lowering drug or to a placebo. As shown in Table 2.3, it was found, after a nine-year follow-up, that 8.1% of the 1,906 men receiving the cholesterol-lowering drug had had a heart attack, as against 9.8% of the 1,900 who took the placebo, for a difference of 1.7%. An article in *Science* stated that these results would "affect profoundly the practice of medicine in this country" (Kolata, 1984, p. 380).

A retrospective, hence non-experimental, study (Goldstein, Andrews, Hall, & Moss, 1992) of the effects of aspirin on heart attacks was done with patients at 15 heart research centers. It found, as shown in Table 2.4, that of those who had not

Table 2.4 Incidence of fatalities from heart attacks in aspirin- and nonaspirin-using patients

Status after 2 years	Aspirin users ($N = 751$)	Aspirin nonusers ($N = 185$)
Dead	1.6%	5.4%
Alive	98.4%	94.6%
Total	100.0%	100.0%

Adapted from "Reduction in Long-Term Cardiac Deaths with Aspirin after a Coronary Event," by R. E. Goldstein et al. (1992). Adapted with permission.

been taking a small daily dose of common aspirin ($N = 185$), 5.4% had a fatal heart attack. Of those ($N = 752$) who had been taking aspirin, only 1.6% had a fatal heart attack. Thus the no-aspirin patients were 3.4 times (5.4/1.6) more likely to have a deadly heart attack than those who had taken aspirin. (Medical researchers often determine the difference between the control-group incidence [%cl and the treatment-group incidence [~°el as a percentage of %c; thus: [%C–%e]/%C. In this aspirin study, that procedure yielded a 70% mortality reduction associated with taking the aspirin. But, of course, this procedure can yield hard-to-interpret results when %c is small; see Weissler, Miller, & Boudoulas, (1989). Lipsey and Wilson (1993) examined the results of 15 meta-analyses of medical-treatment effects on mortality and other medical and psychological outcomes. They concluded that in assessing meta-analytic estimates of the effects of psychological, educational, and behavioral treatment, we cannot arbitrarily dismiss statistically modest values (even 0.10 or 0.20 SDs) as obviously trivial ... [C]omparable numerical values are judged to represent benefits in the medical domain, even when similar outcome variables are at issue. (p. 1199) It might be argued that the medical main effects are taken seriously because the dependent variables are extremely important, often matters of life and death. But many behavioral-science dependent variables – school achievement, dropout rates, mental health, recidivism, occupational adjustment, personal relationships, and group effectiveness – are also clearly important. In any case, apart from the practical importance of the dependent variables, the medical examples suggest that small main effects can have scientific significance – in both medical and behavioral science. The implication is that the results of present-day behavioral science research, in the form of main effects, provide a good basis for behavioral-science theory and practice. The medical experiments and surveys yielded "small" main effects because the treatments worked for some patients but not for others. Why? Presumably because these independent variables interacted with other variables, such as patients' physiological characteristics. Such interactions are not regarded as precluding the possibility of medical science. The present argument applies the same reasoning to uphold the possibility of behavioral science.

Recognizing interaction effects, medical practitioners do not act unthinkingly on a main effect; they use judgment, based on everything else they know about medicine and the patient, including those patient characteristics that might interact with a treatment to produce undesirable effects (see, e.g., Brown, Viscoli, & Horwitz, 1992). Similarly, applications of behavioral science main effects in teaching, counseling, psychotherapy, and business and industry – and in theory development – must depend on judgment, based on everything known about relevant theory, the desired outcomes, the clients, and such other factors as characteristics of settings that might make a given treatment less or more desirable. Medical theoreticians and practitioners do not reject main effects because they do not tell the whole story. Neither should those who use social and psychological main effects in theoretical and practical work. The medical research results cited are taken seriously not only because of their practical value but, of course, because they are supported by relevant theory. The role of theory needs also to be taken into account in appraising the scientific status of the behavioral sciences (Gage, 1994a, 1994b). The present point is that, even with theoretical support, some medical effect-sizes were small. Small behavioral science effects, similarly consistent with theory and other research, should be accorded similar respected. The possibility of science in the presence of interaction effects. Theoretical work needs better information on the generalizability and longevity (i.e., decade × treatment interactions) of the main effects with which behavioral science is concerned. Before meta-analysis, knowledge about the strength, durability, and consistency of main effects rested primarily on the traditional nonquantitative, excessively impressionistic, and unsystematic literature review. Such reviews were additionally flawed by their taking too seriously the statistical significance of the results of single low-power studies. Now it is understood (Gage, 1978, pp. 24–31; Hedges & Olkin, 1988; Schmidt, 1992) that vote-counting conducted on the basis of such statistical significance almost certainly leads to errors of statistical inference whereby a false null hypothesis goes unquestioned. Meta-analysis can reveal when the results of replications vary so widely that we should suspect the influence of additional variables on the main effect. In that event, meta-analysis can identify what variables make how much of a difference in the magnitude of main effects (see, e.g., Cook et al., 1992). It can reveal, across replications, whether the difference between the effects of, for example, two different kinds of teaching is larger or smaller or even reversed in, for example, classes with middle-class students as against classes with low-income-family students and so on. Such results of meta-analysis, which are indeed interaction effects determined across replications, point to qualifications needed in interpreting main effects. Becker (1996) has proposed "a systematic theory of the generalizability of research results, drawing on formal quantitative methods for research synthesis and on the theory of generalizability of measurements" developed by Cronbach, Gleser, Nanda, and Rajaratnam (1972). This approach would mean studying main effects with the same rationales that Cronbach and his co-workers developed for studying test scores. Becker's formulation makes possible the quantitative study of the generalizability of main effects and also of interaction effects. But here, as was noted above, the interaction effects can be studied through meta-analysis, not within single studies but across replications, where the possibly interacting variables that influence the main effects are not only aptitudes, or characteristics of persons, but also variables in the measures and settings used. The results of such generalizability studies may conceivably be incorporated into. Guttman-type facet theorizing with the built-in strengths of replications. High statistical generalizability is more persuasive when the replications meta-analyzed are more heterogeneous in the types of persons, measures, and contexts represented. Thus, the same results obtained in 20 highly similar studies carry less weight for generalizability than such same results in studies whose subjects vary in, for example, ethnicity and educational level, whose measures come from both tests and observations, and whose contexts extend from the laboratory to real-life situations.

In mentioning Thomas's opposition to theory, I find his argument further weakened when we realize that "theory" is synonymous with "explanation." For example, one collection of papers on theory by philosophers of science is titled *Theories of Explanation* (Pitt, 1988). Giving up the search for a theory of teaching is equivalent to giving up the search for explanations of educational phenomena. Although Thomas (2007) is willing to accept and respect theory in the natural sciences, his position against theory for phenomena in education, such as teaching, means that he is willing to give up the search for explanations of phenomena found in teaching. To provide explanations of teaching – explanations of why and how it works and why some kinds of teaching work better than others – is the purpose of this book.

In the natural sciences, the desirability of theory has gone unquestioned for centuries in the writings of philosophers and scientists. Scientists regard valid theory as the ultimate goal of scientific research. Among the achievements of physical scientists have been such triumphs as the theory of motion, electromagnetic theory, relativity theory, quantum theory, and the theory of the chemical bond. The biological sciences point with pride to Darwin's theory of evolution, Pasteur's germ theory of disease, Mendel's genetics, the Watson-Crick double-helix structure of DNA, and much more.

Until recently, the same unanimity about the value of theory prevailed in the social and behavioral sciences, even though their theoretical achievements were less glorious or well-established. In psychology, some examples are the general factor (g) theory of intelligence, attribution theory, classical and operant conditioning theory, construct theory, and the theory of cognitive dissonance. Major volumes (e.g., Koch, 1959–1963) have brought together some of psychology's attempts at theory. In sociology, we find social mobility theory, the theory of group structure, and much more, as well as volumes devoted to the development of theory (e.g., Berger & Zelditch, 1993; Cohen, 1989; Dubin, 1969; Glaser & Strauss, 1967).

Thus, for a long time, to question whether valid theory was desirable was nearly unthinkable. Until the last few decades, natural and social scientists and philosophers regarded valid theory as the pinnacle of scholarly and scientific achievement.

The general and centuries-old conviction among scientists and scholars agreed with the aphorism, "There is nothing so practical as a good theory" (Kurt Lewin, quoted in Marrow, 1969, p. viii). But Lewin's statement calls for the specification of the *practical* values that theory serves. For example, the theories of astronomy – however valid, enlightening, and exquisite as explanations of the location, motion, clustering, and composition of heavenly bodies have value only for the guidance of astronomers engaged in further searches for understanding. Except for use in navigation and meteorology, astronomical theory has little or no *practical* value, in the sense of being useful in the everyday affairs of persons who are not astronomers.

In other natural sciences, theory has indeed had practical value, that is, has served as a basis for the development of technology that meets human needs. Transistors, developed out of solid-state theory, made possible computers, space exploration, and medical tools. Lasers, based on radiation theory, made possible valuable innovations in medical diagnosis and treatment. Magnetic resonance induction theory made possible more detailed examination of the human body's interior. The theory that portrays DNA's double-helix structure made possible the mapping of the human genome and a growing host of practical applications in medicine, agriculture, genealogy, and criminology.

Despite this esteem of theory in science, the philosophy of science, in the manner of all philosophy, consists of "a record of criticism and countercriticism; through such a dialectic philosophers come closer to whatever truth is to be found in their subject matter" (Gewirth, 1991, p. 1). Often, however, those who might be expected to engage themselves with the philosophy of science pay little attention to it. As Scriven (1968) noted,

> Curiously enough... many scientists reject the philosophy of science as irrelevant to their own activities although they constantly talk it and teach it and illustrate its relevance in their own work, sometimes under the title 'methodology' and sometimes just as advice without a label. (p. 84)

Indeed, my own informal survey of three recent recipients of the doctorate in chemistry from major universities indicated that they knew little or nothing about the ideas of important twentieth-century philosophers of science: Paul Feyerabend, Thomas Kuhn, and Karl Popper.

By the same token, a theory of teaching should also yield practical benefits, because teaching is central to the process of education, and education is indispensable to the freedom and well-being of individuals in modern societies. Here is how the physicist Louis Ridenour (1950), made this point at a meeting of the American Educational Research Association:

> The first observation that we can draw from a study of research in the natural sciences, and of the engineering which rests on scientific knowledge, is this: when one is faced with the necessity of solving an important practical problem, he may be forced to attack it directly; but a better solution than that yielded by a direct attack on a problem is likely to come from an unexpected source, usually from an investigation in "pure science" which has been undertaken altogether for its own intellectual interest. That is, the great advances in practical engineering technique have always arisen from random scientific investigations aimed at no more immediate object than that of understanding the world a little better.
>
> A parable due to Ernest Lawrence, Berkeley physicist and Nobel laureate, may serve to illustrate this point. Suppose, says Lawrence, it had been decided in the year 1840 that methods of indoor illumination required improvement. Suppose that the government had established an indoor illumination commission, with plenty of money and contract authority, with great laboratories staffed with competent scientists and engineers. There would have been progress, sure: better designs for lamp chimneys, more efficient fuels than animal oils, improved wicks, and the like. With considerable effort, a two- or three-fold improvement in the economy and effectiveness of indoor illumination might have been produced in a few years time. But it is entirely certain that no one connected with the indoor illumination commission would have spent any time waving wires in front of magnets, connecting together with a wire of metal two dissimilar metal plates immersed in a common electrolyte no one

would have done any of the things which in fact led, before a generation had passed, to our giant and indispensable electrical industry....

In short, it is quite possible to be so "tough-minded" and "practical" as to miss entirely the discoveries that are of major importance in what you need to do. (Reprinted in Gage, 1960)

Hirsch (2002) expressed a similar admonition: Educational research cannot be scientific unless it cultivates and produces theory that will yield causal explanations of the outcomes of educational treatments. A theory of teaching should improve our abilities to *explain* (understand) teaching, *predict* the determiners and consequences of teaching, and *control* (that is, improve) the effectiveness of teaching.

The desirability of a theory of teaching also results from the desirability of improving the competence, attractiveness, and rewards of the teaching profession. After surveying the field of teacher competence and teacher education in the U.S., Kerr (1983) recommended various policy changes, the first of which was that research universities should develop a three-year professional doctoral program in teaching:

a new professional doctorate grounded in theory and professional studies... A doctor of teaching for every 20–30 regular teachers in a school would be both affordable and efficient. The regular teachers would employ the most sophisticated diagnostic measures, instructional techniques, and assessment instruments with the guidance, supervision, and counsel of the head teachers. (pp. 142–143)

The desirability of a theory of teaching was implicit in the statement by Finn (2000) of a view of the need for, *and prospect of* a scientific basis for the art of teaching, as follows:

You will hear it said... that schools of education should be more like schools of medicine and that teaching as a profession should be more like medicine as a profession. There is an important reason, however, why they are very different and likely to stay that way for some time to come. For better or worse, medicine rests on science, on a reasonably stable body of knowledge, based on high-quality, replicable research accepted by everyone in the field and systematically imparted by its training institutions. I look forward to the day when teaching will be that way, too. But it is not that way today, and it may not be for the next two centuries. (pp. 11–12)

Finn's guess as to how long it may take to have a scientific basis, and hence a theory, of the art of teaching cannot be evaluated now. Similarly, no one could have predicted – in the 1840s, when Faraday was waving wires in front of magnets – the emergence in the 1880s of electric power industries.

But, Is Human Teaching Moribund?

Thus far, we have argued the case for theory of teaching. But what is the case for a theory of *teaching done by human beings*? Since at least the 1960s, teaching by nonhumans, i.e., by programmed booklets, machines, and computers, has been developed and promulgated. At first, "programmed instruction," developed by B. F. Skinner (1957) and brought together in a large volume of writings edited by Lumsdaine and Glaser (1960) caused concern that "a specter is haunting research on teaching the specter of programmed instruction" (Gage & Unruh, 1967). That vision was expressed again by Adams (1971):

> The suggestion has been made elsewhere that classroom research, in the face of the "specter" of automated education, is the ultimate gesture in futility. After all, what virtue can inhere in researching a phenomenon that tomorrow will be as dead as a dodo? (p. 101)

Writing a theory of teaching early in the twenty-first century might have the same fate as that of a theory of medicine written in the years (about the 1850s) before Pasteur's germ theory of disease revolutionized medicine. Adams offered a possible consolation: Even if human teaching disappeared, we should "redouble our efforts – not indeed in order to solve pedagogical problems, but so that a quaint form of twentieth century ritual can be preserved for posterity's interest, edification, and amusement (p. 101)."

The present effort rests on a different vision – that nonhuman teaching will not replace human teaching. Rather, it has been a gradual development characterized by teachers' learning how to use computers and their programs in somewhat the same way in which they learned to use textbooks, tests, chalkboards, libraries, laboratories, workbooks, slides, films, portfolios, and, more recently, digital video displays (DVDs). Despite all these additions to the teacher's tools, the human teacher has retained her central role in education. Cuban (1993) convincingly documented the ways in which all these innovations, including computers, could be summed up with a headline: "Computers Meet Classroom: Classroom Wins."

The present effort rests on a different vision – that nonhuman teaching will not replace human teaching. Although nonhuman teaching has become more prevalent since the 1960s, its gains have not replaced teachers (Cuban, 2001). Its growth has not had the pace of a revolution. As Cuban put it, after a thorough investigation of many aspects of the issue, "When it comes to higher teacher and student productivity and a transformation of teaching and learning, however, there is little ambiguity. *Both must be tagged as failures.* [italics added] Computers have been oversold and underused, at least for now" (p. 178).

Conceptions of Theory

There are many usages of the term "theory," and they vary widely. Almost all conceivable implicit definitions have appeared in one place or another at one time or another. Chambers (1992, pp. 7–27) identified fifteen usages, which he put into nine clusters: (1) Theory as contrasted with fact; (2) Theory as contrasted with practice; (3) Theory as evolving explanation; (4) Practical theory that guides a profession or art; (5) Theory as hypothesis; (6) Theory as ontological or observational presupposition; (7) Normative theory, such as doctrine or dogma; (8) Empiricist theory; (9) Scientific theory.

Of these, only the last two – empiricist theory and scientific theory – are considered here. An empiricist theory, according to Chambers (1992), is one in which generalizations about *observable* variables are related to one another in ways that accord with empirical observations. A scientific theory, Chambers (1992) wrote, is one that relates *abstract* (e.g., mathematically or logically manipulable, but not observable) concepts and their variables in logical or rational ways that accord with empirical observations.

Chambers obtained his conception of science from the sociologists Willer and Willer (1973), who referred to two types of thought – theoretical and observational – as shown in Fig. 2.1.

The observational level deals only with empirical categories, and the theoretical level deals only with abstract concepts. Connections on only the observational level are empirical connections, those on only the theoretical level are rational connections, and connections between the empirical and abstract levels are *abstractive*. Thus the Willers (and Chambers) held that science and empiricism are both concerned with observed empirical events, but "The logical form of science is much more complex" (Willer & Willer, 1973, p. 15).

Scientific knowledge, then, consists not only of successful rational connections between concepts, not only of successful observational connections, but also of abstractive connections, which are rational connections that correspond to observational connections.

The Willer-Chambers conception of theory may have stemmed from assumptions that only the theories of the physical sciences are genuine theories and that only mathematically expressed theories were to be considered theories. Chambers used as examples only theories from the *physical* sciences: Copernicus's classical mechanics, Kepler's laws of planetary motion, Galileo's laws of motion, Lavoisier's role of gases in chemical reactions, Crick's and Watson's structure of the DNA molecule, and Wegener's continental drift. Chambers apparently considered claims to theory outside the physical sciences to represent mere pretension.

This conception of theory is not held by all scientists and philosophers of science. Thus the philosopher of science Nagel (1979) wrote that

> The requirements for being a genuine science tacitly assumed in most of the challenges [to the scientific status of the social sciences] lead to the unenlightening result that apparently none but a few branches of physical inquiry merit the honorific designation. (p. 449)

But, said Nagel, despite their limitations,

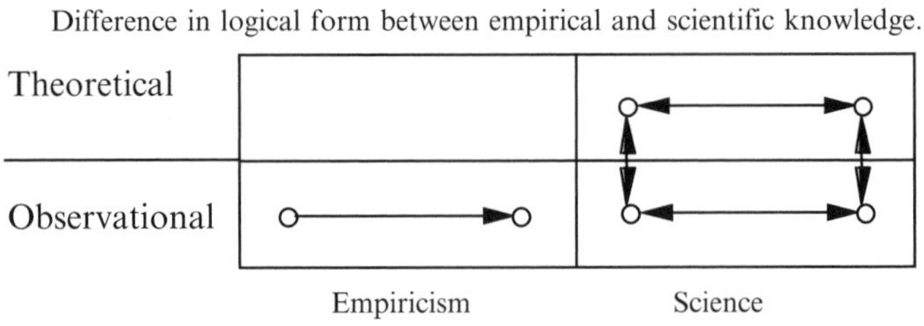

Difference in logical form between empirical and scientific knowledge.

Fig. 2.1 Difference in logical form between empirical and scientific knowledge (Source: Willer & Willer, 1973, p. 19).

the generalizations of social inquiry do not appear to differ radically from generalizations currently advanced in domains usually regarded as unquestionably respectable subdivisions of natural science – for example, in the study of turbulence phenomena and in embryology. (p. 449)

In any case, our present goal is to set forth a theory of teaching that will explain in logical, or intuitively reasonable, terms the empirical relationships between important concepts, or variables, that characterize teaching.

As we noted above, one widely held conception of theory in both the natural and social sciences is that theory must *explain* the empirical phenomena, such as the relationships between concepts and variables. Examples of relationships in psychology would be explanations of (a) the universally found positive correlation between the socioeconomic status of individuals and their cognitive ability, and (b) the higher correlation between the IQs of identical twins than that between the IQs of fraternal twins.

But, because "explanation" can mean many things, it needs to be defined. Accordingly, a substantial literature on theories of explanation, written mostly by philosophers of science, has appeared (see e.g., Pitt, 1988). And in that literature, one widely accepted conception of explanation is the "covering law explanation" (Hempel & Oppenheim, 1948). Their first example of such an explanation is that of a mercury thermometer rapidly immersed in hot water.

There occurs a temporary *drop* [italics added] of the mercury column, which is then followed by a quick rise. How is this phenomenon to be explained? The increase in temperature affects at first only the glass tube of the thermometer; it expands and thus provides a larger space for the mercury inside, whose surface therefore drops. As soon as by heat conduction the rise in temperature reaches the mercury, however, the latter expands, and as its coefficient of expansion is considerably larger than that of glass, a rise of mercury level results. This account consists of statements of two kinds. Those of the first kind indicate certain conditions which are realized prior to, or at the same time as, the phenomenon to be explained; we shall refer to them briefly as antecedent conditions. In our illustration, the antecedents include, among others, the fact that the mercury thermometer consists of a glass tube which is partly filled with mercury, and that it is immersed into hot water. The statements of the second kind express certain general laws; in our case, these include the laws of the thermic expansion of mercury and of glass, and a statement about the small thermic conductivity of glass. The two sets of statements, if adequately formulated, explain the phenomenon under consideration; they entail the consequence that the mercury will first drop, then rise. *Thus the event under discussion is explained by subsuming it under [covering it with] general laws, that is, by showing that it occurred in accordance with those laws, in virtue of the realization of certain specified antecedent conditions.* (Pitt, 1988, pp. 9–10)

Accordingly, a theory of teaching should explain how it is that students learn from teaching and do so by invoking more general "covering laws" of human behavior.

Must Scientific Research Be Theory-Driven?

In relation to scientific research, the value of theory can be differentiated between (a) theory formulated and used *before* the collection and analysis of empirical observations, i.e., "theory-driven research," and (b) theory produced *after* the collection and analysis of observations.

The Prior-Theory-is-Indispensable Position

One argument in favor of theory is that *prior* theory is a necessary tool of scientific research. Some writers have gone so far as to claim that nothing scientific can be learned from research conducted without a *prior* theory that drove the research.

Lewin (1931), the protagonist of "field theory" in psychology, labeled as "Galilean" the theory-driven conception of scientific research, because he held Galileo to have been a prime exemplar of that approach. In contrast, his "Aristotelian" science – the science of Aristotle as interpreted by Thomas Aquinas (Martin & Sugarman, 1993, p. 18) – sought generalizations derived empirically on the basis of no prior theory. "Aristotle viewed scientific inquiry as a progression from observations to general principles and back to observations" (Losee, 1980, p. 6). Akin to Galilean principles, Garrison and Macmillan (1984) stated that

> It is only when the research is theory-driven from within that the theories themselves are supported or falsified; and only when this is done can there be better theories for the explanation and direction of pedagogical practice. (p. 3)

Thus these writers claim that a theory is necessary at the beginning, before observations are collected, analyzed, and interpreted. Similarly, Chambers (1989) held that

> [I]mprovement of [scientific] enterprises does not derive from the accumulation of further facts. It derives from the introduction of theory of various types which can be used both to say what count as facts, to make sense out of them, and to test them in various ways. (p. 84) [U]ntil very recently, Process-Outcome researchers [on teaching] do appear to have been unaware of the manifold significance of theory... in science ... They are, for example, still largely unaware of the effects of theory on what is observed, and thus on what can count as a fact. (p. 86)

The same position was taken, somewhat vehemently, by Ball (1995) in regard to educational research:

> I wish to argue that the absence of theory leaves the researcher prey to unexamined, unreflexive preconceptions and dangerously naive ontological and epistemological *a prioris*. I shall wail and curse at the absence of theory and argue for theory as a way of saving educational studies from itself. (pp. 365–366)

Martin and Sugarman (1993) described the distinction between Aristotelian and Galilean approaches to scientific research as a contrast between the Aristotelian focus on *observed regularities* and the Galilean focus on constructing *hypothetical models* of underlying realities. Aristotelian forms of scientific thought focus on identifying the essential qualities of objects associated. Aristotelian science discovers concepts inductively, from analysis of observations, while Galilean science invents concepts and relies on prior theory.

The Prior -Theory-is -Not-Indispensable Position

The alternative position is that, although valid theory is the supremely desirable *outcome* of scientific research, *prior* theory is not indispensable. In support of that position, Gage (1994b) pointed to the problems that the "prior-theory-is-indispensable" position runs into in the light of (a) the concept of implicit theory, (b) the positions of some philosophers of science, (c) the serendipity, or *theoryless* research, that has led to important scientific discoveries, and (d) the history of research on teaching.

Implicit Guiding Theories

The first reply refers to the idea of *implicit guiding theory*. The critics apparently reject the possibility that scientific research can proceed without prior formulation of an explicit theory and its associated hypotheses. No matter how well what the scientist does satisfies other ideals of scientific method (maximized rationality and objectivity, precision of definitions, public character, replicability, and falsifiability), if the prior theory has not been explicitly formulated, say these critics, the whole enterprise cannot be considered scientific.

It seems that scientific research, in the view of these critics, cannot evolve through a process moving from implicit, ill-defined, perhaps *ad hoc,* and exploratory, but nonetheless influential theory, and not requiring prior theories that are explicitly formulated, and then survive attempts to falsify the theories with empirical evidence. The critics used no detailed knowledge of such relatively early research on teaching as that of Mitzel (1960), Flanders (1970), and Medley (1977), among others, who did process-outcome research on teaching before the 1980s. They also did not look carefully at the more recent process-outcome research on teaching. So they seem unfamiliar with the reasoning (that is, the implicit theory), either openly stated or readily inferred, underlying the researchers' choice of concepts, variables, and measuring instruments, and thus the easily inferred implicit theories that these researchers were testing. In short, they do not seem to have benefited from the admonition by the philosopher of science Hanson (1958, p. 3): "Profitable philosophical discussion of any science depends on a thorough familiarity with its history and its present state."

Marland (1995) examined the concept of implicit theory from the standpoints of the student, the teacher in training, the practicing teacher, and the teacher educator. All of these use implicit theories of teaching in the form of *metaphors*: the teacher as "cook, entertainer, counselor, timekeeper, engineer, preacher, conductor, mother figure, horticulturist, actor, and ship's captain" (p. 134).

More such metaphors come readily to mind: the teacher as explainer, chairperson, prosecuting attorney, and even

> "The teacher as a Bayesian sheep dog." The resulting image is of a barking collie propelling his bulging flock along a path by successive statistical estimation and adjustment of the flock's average direction, while racing to keep diverging individuals contained with the group. (Snow, 1973, p. 89)

(Bayes's principle is that "the probability of a given event is a consequence of a specified one of a number of mutually exclusive antecedent conditions which might have given rise to the event" [English & English, 1958, p. 51]). Similarly, the "Bayesian approach in evaluation research [is] the use of conditional probabilities as an aid in selecting between various program outcomes." (VandenBos, 2007, p. 105).

Implicit theories can also, said Marland (1995), take the form of *images* whereby the teacher or researcher might see "the classroom as a home" (Clandinin, 1986) with all of the home's attendant feelings and emotions.

In considering research on teaching that has not been *explicitly* theory-driven to be *ipso facto* unscientific, the critics ignored the strong probability that it has been driven by *implicit theories*. Whenever investigators choose variables and develop ways of measuring them, they operate on at least an implicit theory if not an explicit one. The philosopher of science Hanson (1958) had such a conception of implicit theory in mind when he wrote that

> There is a sense, then, in which seeing is a "theory-laden undertaking." Observation of x is shaped by prior knowledge of x. Another influence on observations rests in the language or notation used to express what we know, and without which there would be little we could recognize as knowledge. (p. 19)

So it follows not only that the investigator must have at least an implicit theory so as to be able to carry out the necessary research steps in a nonrandom way but also that *atheoretical* educational science is impossible, as was asserted by Garrison and Macmillan(1987). Researchers may not be aware of their theory, but they act on it nonetheless.

Such implicit theory is easily seen in the dimensions of teaching specified in the classroom-interaction-analysis categories used by Flanders (1970); those categories reflect the implicit theory that classroom climate along a direct-indirect influence dimension is related to student achievement and attitude. Such implicit theory is also evident in the dimensions investigated by Brophy and Evertson (1976) – dimensions that were based on the readily inferred theory that classroom management, that is, ways of holding student attention and minimizing academically counterproductive use of time, was related to student achievement. A third example is the widely applied instrument developed by Stallings (1975) to reveal the degree of implementation of ten different "theoretical" models of teaching developed as planned variations in Project Follow Through; each model had been developed and rationalized by a separate team.

To anyone who looks into the background of the observation-of-process schemes used in process-outcome research on teaching, their underlying implicit theories are evident even though they vary in the explicitness with which they are stated. The assumption that process-outcome research has been atheoretical seems to result from failure to examine these authors' writings and their process-observation instruments. In none of the philosophical criticisms is there any evidence of a careful reading of the details of the process-outcome research reports. It is as if, unless researchers used the word "theory," these critics were unable to see the theory.

Philosophers of Science on Scientific Method. A second reply refers to what philosophers of science have written on the idea of fixed and firm conceptions of scientific method. Much could be found in their writings to raise questions about the rigid notions of these critics concerning the nature of scientific method. Here are a few such statements, beginning with that of Peter Medawar, not because he was a philosopher but because he was a Nobel laureate who was quoted by one critic (Chambers, 1989, pp. 83–84) as an authority on the nature of science. Chambers ignored Medawar's statement that

> There is indeed no such thing as "the" scientific method. A scientist uses a very great variety of exploratory stratagems, and although a scientist has a certain address to his problems – a certain way of going about things that is more likely to bring success than the gropings of an amateur – he uses no procedure of discovery that can be logically scripted. (Medawar, 1984, p. 51)

The same kind of insistence against any rigid codification of the methods to be used by scientists occurs in the writings of both "radical" and "nonradical" philosophers of science. Thus the "radical" philosopher of science Feyerabend (1963, p.13) argued that science proceeds according to no rational method. In his view, scientists, in making progress, have violated every logical principle and followed the motto "anything goes."

Similarly, the "nonradical" philosopher of science Nagel (1977) stated that the notion of *the* scientific method is so weak that it is arguing against "a straw man" to argue *against*

> a notion of method according to which there are firm, unchanging, and absolutely binding principles for conducting the business of science. It would be difficult to find many contemporary scientists or philosophers of science who hold that the principles they may be employing in assessing the validity of cognitive claims in science cannot be modified and improved in the light of continuing theory. (p. 71)

These statements mean the opposite of what the philosophical critics of research on teaching have insisted upon, namely, the inescapable requirement that scientific research be explicitly theory-driven.

These critics have also ignored the possibility of replicating a study, *using as a basis for a hypothesis a finding of a first study.* Such replication with theory on the basis of results that were not previously explicitly hypothesized has occurred frequently in the history of science and in process-outcome research on teaching. The atheoretical first study's finding leads to a theory confirmed or disconfirmed in a subsequent study.

Serendipity in the History of Science. A third reply deals with empirical evidence on the necessity of prior theory in scientific research. That evidence is found in the history of science – a history that reveals many important exceptions to the argument that scientific research must be explicitly theory-driven.

Especially noteworthy here is the literature on *serendipity*. The critics of research have ignored the long and important history of the role of serendipity in scientific research (see, e.g., Kanterovich & Néeman, 1989). That history demonstrates that extremely important scientific discoveries have been made without any prior theory or hypothesis. Among such discoveries are X-rays, the electricity-magnetism connection, penicillin, cosmic microwave background radiation, and pulsars. (See Table 2.5.)

Table 2.5 Bibliography on serendipity

Baggott, J. (1990). Serendipity and scientific progress. *New Scientist*, 67–68.

Beveridge, W. I. B. (1957). *The art of scientific investigation* (3rd ed., Ch. 3). New York: Vantage.

Cannon, W. B. (1940). The role of chance in discovery. *Scientific Monthly, 50,* 204–209.

Kanterovich, A., & Néeman, Y. (1989). Serendipity as a source of evolutionary progress in science. *Studies of History and Philosophy of Science, 20*, 505–529.

Koestler, A. (1964). *The act of creation.* New York: Macmillan.

Mach, E. (1896). On the part played by accident in invention and discovery. *Monist, 6,* 161–175.

Roberts, R. M. (1989). *Serendipity.* New York: Wiley.

Shapiro, G. (1986). *A skeleton in the darkroom: Stories of serendipity in science.* New York: Harper & Row.

Simonton, D. K. (1989). Chance-configuration theory of scientific creativity. In Gholson, B. (Ed.), *Psychology of Science: Contributions to meta-science.* New York: Cambridge University Press.

Source: Kanterovich and Néeman (1989).

It might be argued that even serendipity is theory-driven because the researcher must have a theory in order to recognize the anomalous character of the serendipitous finding and to appreciate its deviation from expectations. If so, any such recognition of the theory-driven character of even serendipitous findings leads us again to accept the concept of implicit theory: theory that is guiding the research despite the fact that the theory has not been clearly, distinctly, and explicitly stated. The researchers would not have formulated the implicit theory *before* their serendipitous finding. Rather, the researchers recognized that the finding violated their initial, unstated, and implicit theory and called for a reformulation of either an explicit or an implicit theory.

Wilhelm Roentgen, for example, had not stated beforehand any explicit theory that certain rays could pass through opaque matter and affect a photographic plate. It was only after he serendipitously observed such an event in 1895 that he worked on that theory. He called them X-rays.

Similarly, while working with Staphylococcus bacteria in 1928, Alexander Fleming serendipitously noticed a bacteria-free circle around a mold growth that was contaminating a culture of staphylococci. Upon investigating, he found a substance in the mold that prevented growth of the bacteria even when diluted 800 times. He called it penicillin.

Some History of Process-Outcome Research on Teaching. A fourth reply refers to the history of process-outcome research on teaching. How do the ideas just mentioned, from the philosophy and history of science, bear upon the history of process-outcome research on teaching?

The critics did not look carefully at the history and origins of process-outcome research on teaching. They thus deprived their thinking, about its scientific status, of information about where process-outcome research on teaching came from and, especially, why it developed.

How research on teaching has progressed since the years before the 1950s can be seen in the variables whose research records were examined in extensive reviews of earlier research on teaching (Domas & Tiedeman, 1950; Morsh & Wilder, 1954; Getzels & Jackson, 1963). Those variables are now called "presage" variables (Mitzel, 1960) as distinguished from "process" variables. Presage variables deal with the teacher's personality and characteristics, such as intelligence level, educational level achieved, academic success, age, years of experience, knowledge of subject matter, extracurricular activities, general culture, socioeconomic status, sex, marital status, performance on paper-and-pencil tests intended to measure "teaching aptitude," inventories of attitude toward teaching, voice and speech characteristics, appearance, and personality characteristics as measured with rating scales and questionnaires. By and large, these presage variables turned out – as summarized in the reviews cited above – to be unrelated to student achievement.

But more recent studies have found significant relationships between teacher personality and characteristics and measures of student achievement. Nonetheless these reviews of the literature in the 1950s and '60s still explain the shift, beginning in the 1960s, from presage to process variables. Hindsight makes it easy now to consider those early efforts misguided and doomed to failure in discovering correlates of student achievement. In any event, the record of that research sufficed to make investigators (beginning in the 1960s) turn away from presage to process variables. The investigators who conducted the studies of presage variables presumably based their work on at least implicit theories and hypotheses; some of them stated those hypotheses and their rationales in their reports. Nonetheless, the work on the teacher's personality and characteristics, at least as formulated in those studies, did not pay off.

So it was that Medley and Mitzel (1963) wrote what turned out to be an imperative heeded by many subsequent researchers on teaching. They said:

> Certainly there is no more obvious approach to research on teaching than direct observation of the behavior of teachers while they teach and pupils while they learn. Yet it is a rare study indeed that includes any formal observation at all. In a typical example of research on teaching [before 1963], the research worker limits himself to the manipulation or study of antecedents and consequents of whatever happens in the classrooms while the teaching itself is going on, but never once looks into the classroom to see how the teacher actually teaches or how the pupils actually learn. (p. 247)

Accordingly, the subsequent process-outcome work did not arise from the thoughtless, theory-free thrashing-about seemingly imputed to such studies by the critics. It arose from discouraging experience with presage variables, from the realization that better results might come from looking at what happens in the classroom, and from careful thought about what variables in classroom process might make a difference in what students learned. Only a careful reading of process-outcome research reports would permit anyone to judge the degree to which the research was theory-laden.

Karl Popper's Resolution of the Issue

The philosopher of science Karl Popper (1963) spoke to the issue of whether all scientific research must be theory-driven by substituting the concept of *conjecture*[1*] for "theory." The conjecture should, of course, stem from the best evidence and logic available to the investigator. A conjecture claims much less than a theory; *it is an assumption, a guess, a hunch, a speculation, a supposition, a surmise, a jumped-to-conclusion.*

Popper held that inductive logic cannot serve as a way to *prove* the truth of a conclusion. Even a phenomenon, such as the setting of the sun in the west, that has been observed to occur predictably in the same way, without exception, innumerable times, over many millennia, cannot prove, *by inductive logic alone,* without the support of deductive-logical principles, that the sun will always set in the west. As Popper (1963) told it,

> Thus I was led by purely logical considerations to replace the psychological theory of induction by the following view. Without waiting, passively, for repetitions to impress or impose regularities upon us, we actively try to impose regularities upon the world. We try to discover similarities in it, and to interpret it in terms of laws invented by us. Without waiting for premises we jump to conclusions. These may have to be discarded later should observation show that they are wrong.

This was a theory of trial and error – of conjectures and refutations. It made it possible to understand why our attempts to force interpretations upon the world were logically prior to the observation of similarities. (p. 60)

Popper's phrases – "try to impose regularities upon the world ... to discover similarities in it to interpret it in terms of laws invented by us ... jump to conclusions ... force interpretations upon the world that were logically prior" – are Popper's modest equivalents of "theory." But, whereas "theory" connotes formality and rigor, "conjecture" has an informal and fallible flavor. Indeed, Popper's conception of scientific method, stated in his book's title, *Conjectures and Refutations,* expresses something similar to the notion of implicit theory – a conjecture to be subjected to strong efforts at refutation and to be accepted only as long as it survives those efforts.

An Empirical Approach to Controversies About Scientific Method

The arguments on both sides of the Aristotelian versus Galilean issue seem eminently reasonable. Philosophers and scientists hold strong opinions on the basis of impressionistic and case-study evidence. But empirical research on

---

[1*] "The formation or expression of an opinion or theory without sufficient evidence or proof." (*Random House, Webster's College Dictionary*, 1991).

scientific methods and their outcomes might show which side is more valid in terms of its results.

The Faust-Meehl Proposal

Consequently, Faust and Meehl (1992) proposed that scientific method be used to resolve questions in the history and philosophy of science. Such an effort would entail (a) identifying representative samples of scientific projects in a domain, (b) defining theoretically promising characteristics of the projects, such as whether or not they were theory-driven, (c) measuring the degree to which the projects manifested these characteristics, (d) evaluating those projects as to the scientific value of their yield, and then (e) determining the degree to which the various characteristics correlated with the scientific value of their outcomes and actually differentiated between the projects on the dimension of scientific value. The sample selection, the evaluation of yield, and the identification of the characteristics of the projects represented in their samples, are feasible undertakings, especially in relation to the importance of the issues to be resolved.

In any event, the Faust-Meehl proposal sharpens appreciation of the logical invalidity of the Chambers (1992) approach to the determination of the source and content of the abstract concepts – their discovery and manipulation – in his conception of science. As already noted, even if research on the characteristics of successful physical science yielded intellectually significant findings, we would want, of course, to know whether those findings hold for the biological, social, and behavioral sciences, all of which Chambers ignored in displaying examples of science. Further, Chambers used what might be called his "clinical judgment" in appraising both the track record and the promise of approaches to research on teaching. But, as Faust and Meehl (1992) remarked, such judgment has, with great consistency, had much less success, as compared with actuarial methods, in predicting all sorts of outcomes. The huge body of research on clinical versus actuarial judgment convincingly shows that even crude, non-optimized decision procedures that combine information in a linear manner consistently equal or exceed the accuracy of human judges (Faust & Meehl, 1992, pp. 197–198).

The Faust-Meehl proposal could be used for improving prediction of the scientific payoff of various approaches to doing science, including such approaches as formulating abstract concepts without being influenced by observations, which Chambers (1992) sees as the method of great physical scientists. Even if the Faust-Meehl proposal is never carried out, it signals the untrustworthiness on these matters of strong opinions based on inadequate empirical evidence.

Empirical Consensus in Defining Science

A second consequence of using empirical facts in the history of science in defining science is that those empirical facts include the ways in which other philosophers and historians of science have used the term *science*. Thus one is led to ask, What is the consensus among such thinkers?

An example of how a philosopher of science regards the scientific status of the social sciences is provided by Glymour (1983). He first dismissed the notion that "all really scientific explanations follow the pattern of a sort of logicized celestial mechanics." Then he asserted that

> Social sciences are in large part applied sciences, closely tied to our wish to know the effects of social policies or the causes of social phenomena. Much of the work in contemporary sociology, educational research, political science, and econometrics is directed towards such ends, and what it produces are *causal explanations* ... of particular social phenomena, and descriptions of the *causal relations* among social phenomena. By and large, the explanations and descriptions produced are of a kind, and are buttressed by the sort of arguments that we recognize as scientifically rational, and their content is often useful and surprising. Social scientific work of this kind produces explanations, causal explanations; and knowledge, causal knowledge, without producing general laws, at least not the sort of general laws the critics of social science demand. In doing as much, social science follows a pattern that is common throughout the sciences. It is a pattern most common in applied sciences, in epidemiology, in biology, and in engineering. It is least common, but scarcely absent, in physics. (pp. 127–128)

Thus it is far from a unanimous consensus among philosophers of science that the only correct proper and honest usage of the term "science", one that is free of kudos-seeking pretentiousness, is one congruent with the physical sciences. In this empirical manner, parallel to that of Chambers (1992) in drawing upon the history of the physical sciences, we find that the social and behavioral sciences are sciences.

Theory as the Outcome of Research

So theory is not an absolute *prerequisite* of scientific research. But valid theory does gain absolutely high status as an outcome, the *summum bonum*, of scientific research. Philosophers of science agree on the desirability of valid theory as evidence that scientific research has succeeded. The prominent philosopher of science who decried theory as the goal of science was Paul Feyerabend (1993), whose ideas are considered below.

The Neglect of Theory in Educational Research

During the decades after World War II, American federal research agencies – such as the National Science Foundation and the National Institutes of Health – were established. Those agencies supported quests for theory in the physical and biological sciences on the well-founded assumption that practical applications of scientific theory would eventually contribute to society's welfare.

In empirical *educational* research in the U.S., however, the question of theory's desirability was largely ignored for a century after the organization in 1867 of the U. S. Bureau of Education – an entity that later became the U.S. Office of Education and then the U.S. Department of Education. In all its forms, that entity paid little attention to theory as a goal, or as part of the method, of inquiry in education.

The neglect of explicit concern with theory has continued in present-day governmental educational research agencies, in non-governmental educational research programs, in the American Educational Research Association, and in state educational research organizations. Only in various foundations that support educational research has work on theory been supported.

This neglect may have stemmed from a fallacious assumption that theory, almost by definition, had no practical value. Because educational research was supported and pursued out of an expectation that it would lead to improvements in education, theory seen as the opposite of practice may have been regarded as a luxury to be deferred until the urgent practical needs of education were met. Or theory may have been regarded as something not to be sought directly but as something that would emerge inductively, even automatically, after sufficiently large accumulations of the results of empirical investigations.

Calls for Theory in Educational Research

In recent decades, calls for theory in educational research have been frequent. In his presidential address to the American Educational Research Association, the philosopher of science Patrick Suppes (1974) offered five kinds of argument:

1. *Analogy.* "The obvious and universally recognized importance of theory in the more mature sciences, e.g., economics and psychology" can be analogized to the "importance of theory in educational research." Adequate theory is generally regarded as the basis for the distinction between empiricism and science.
2. *Reorganization of experience.* Theory "changes our perspective on what is important and what is superficial," as when the obvious (the sun revolves around the earth) is replaced by the nonobvious truth (the earth revolves around the sun).
3. *Recognizing complexity.* Seeking theory leads to investigating *how* and *why* one method of, say, teaching reading is better than another beyond the merely empirical determination of the difference in their effectiveness.
4. *Problem solving.* By comparison with Deweyan problem solving, theory makes it unnecessary to "begin afresh" whenever one is confronted with new problem.
5. *The triviality of bare empiricism.* The mere recording of facts without generalization or theory operates so that there can be no transfer of understanding from one situation or problem to another.

Students of research on teaching have often noted the lack of theory as an outcome. "Theoretical impoverishment" was the term used by Dunkin and Biddle

(1974) in characterizing early research in the field. In their "recommendations for researchers" they held that

> Perhaps the greatest problem facing this field at the moment is our lack of adequate theories of teaching that would integrate and explain its many findings. Concepts this field has, instruments too, and findings by the score. But what do these findings *mean*? Why is it that a given teaching strategy appears in one type of classroom and not another, and *why* does it work whereas other strategies are found to be less effective? Until adequate, empirically based theories are developed, this field will continue to exhibit a complex and somewhat chaotic visage. (p. 425)

Dunkin and Biddle (1974) pointed to a "lack of adequate theories of teaching that would integrate and explain its major findings" (p.425). They also called for "the development of concepts for expressing the processes of teaching" (p. 428), integrating concepts into conceptual systems, and involving the concepts in "explanatory theories that are supported with empirical evidence" (p. 429).

They were joined by Winne (1982, p. 14): "[T]here is much consensus that an essential next step for the field's advance is to generate and test theories." Another demand for theory as a product of research on teaching was made by Shulman (1986a). First, he reviewed the results of process-outcome research, which he considered "easily the most vigorous and productive of the programs of research on teaching during the past [1975–1985] decade" (p. 9). Then he observed that such research had declined in vigor because of its "unabashedly empirical and non-theoretical tenor" (p. 13). It had not yielded explanations of the reasons for the occurrences of certain relationships between what teachers did and what students learned.

Carroll's Model of School Learning

Carroll (1963) expressed a further call for a theory of teaching in his influential "model of school learning" – a model to be described in Chap. 8. His model proposed five basic classes of variables that would account for variations in student achievement. Three of these components were expressed in terms of time. One of the two "non-time" variables in Carroll's model, which he also assumed to be related to achievement, was *Quality of Instruction*. Thus, "The model is not very specific about the characteristics of high quality of instruction" (Carroll, 1989, p. 26). Here Carroll was implicitly affirming the need for theory of teaching. To the extent that instruction is less than optimal, time needed for learning is increased. His model mentions that teachers must (a) tell students clearly what they are to learn, (b) put students into adequate contact with learning materials, and (c) carefully plan and order students' steps in learning.

In the present context, it is noteworthy that Carroll defined quality of instruction only briefly, and his statement provides an implicit argument for the desirability of a theory of teaching. A quarter-century later, Carroll (1989) reaffirmed his position:

Perhaps because the Carroll model of school learning does not deal extensively with elements involved in quality of instruction, it has not been particularly influential in these studies [of quality of instruction]. (p. 29)

That the Carroll model should deal with quality of instruction more "extensively," perhaps through a theory of teaching, is a logical inference from the state of affairs that Carroll described.

Questionings of the Value of Theory

Nonetheless, we should examine recently raised general and specific-to-teaching questions about the desirability of theory – in the natural sciences, in literary studies, in social science, in education as a whole, and particularly in research on teaching.

Alongside the affirmative literature on the desirability of theory, especially theory of teaching, we find other writers who give the impression that they see little or no value in theory.

B. F. Skinner's Position on Theory

In psychology, B. F. Skinner (1938, 1950) seemed to raise doubts about the desirability of theory with the title of his article: "Are Theories of Learning Necessary?" On closer inspection, however, those doubts reflect a misinterpretation of Skinner. The kinds of theory whose value Skinner questioned were those that sought "any explanation of an observed fact which appeals to events taking place somewhere else, at some other level of observation, described in different terms, and measured, if at all, in different dimensions" (1938, p. 193). For example, he questioned theories of behavior based on neurological events, perhaps because, at the time he wrote, knowledge about neurological events was typically metaphorical and could not support any rigorous derivations of behavioral phenomena. In the study of behavior, Skinner was questioning whether such theories – neurophysiological, endocrinological, mental, or conceptual – are valid or necessary for theorizing about behavior.

> The very notion of a neurological correlate implies what I am here contending – that there are two independent subject matters (behavior and the nervous system) which must have their own techniques and methods and yield their own respective observations. No amount of information about the second will explain the first or bring order into it without the direct analytical treatment represented by a science of behavior. (p. 423)

Such attempts to explain phenomena at one level in terms of another, presumably more basic, level are objectionable, said Skinner, because they attempt to explain observed events by appealing to events "taking place somewhere else, at some other level of observation, described in different terms, and measured, if at all, in different dimensions" (Skinner, 1964, p. 385). In Skinner's view, attributing behavior to a neural or mental event, real or conceptual, tended (a) to make us forget that we still need to account for the neural or mental event, real or conceptual, because such attributions were, at least in Skinner's day, unobservable neural and cognitive explanations, and gave us unwarranted satisfaction with the state of our knowledge.

But Skinner raised no questions about the desirability of theories of behavior – theories that explained behavior of one kind in terms of connections between observable behavior and observable stimuli. Indeed, he devoted his career to the development of such theory. And he presumably would have had no objection to cognitive or neural theory that stayed at the mental or neural level, without attributing mental or neural events to other levels.

Levels of Theory

Theories can deal with phenomena at different levels of natural science: physics, chemistry, neurology, behavioral, and mental. Theories at the level of physics deal with such concepts as force, mass, and atomic particles (e.g., electrons, protons, neutrons, neutrinos, positrons) in ways needed to describe and explain certain phenomena, such as X-rays or the motions of physical bodies. At the level of chemistry, theory refers to elements (e.g., oxygen, chlorine) and compounds (e.g., acetylcholine) needed to explain such chemical processes as digestion, cardiac rhythm, and brain activity. At the level of neurology, theories deal with the neural structures (e.g., synapses) and pathways (e.g., afferent and efferent) involved in sensing, perceiving, and remembering. At the level of behavior, theories refer to the observable (visible, audible) actions, such as those involved in responding, approaching or withdrawing, positive and negative conditioning. Finally, theories at the mental-process level use such concepts as working-memory and long-term memory (Atkinson & Shiffrin, 1968), cognitive dissonance (Festinger, 1957), and cognitive load (Sweller, 1999), all involved in learning, comprehending, problem-solving, and teaching.

Reductionism

Skinner's position, although not anti-theoretical, was anti-reductionist. He can be challenged by the success of reductionism in science – explaining phenomena at one level in terms of events at another level. Reductionism has long been fruitful in the biological and physical sciences. In genetics, explanations of secondary characteristics, such as eye color and skin color, have been reduced to, first, chromosomal differences, then differences in genes, and ultimately to differences in chemical processes involving DNA. In physics, matter was reduced to compounds, then elements, and subsequently, particles such as protons, electrons, and neutrons.

Psychologists have reduced social processes to their psychological bases. For example, we can understand free enterprise in part, by referring to the socio-psychological process of competition. Some psychological processes have, in part, been reduced to their neurological and glandular bases; for example, the experience of anger results from activation of the adrenal gland by neural processes in the brain, and those neural messages may be explained in chemical terms, such as the flow of the chemical compound acetylcholine. Some mental processes occur at the behavioral level, describing aspects of observable behavior (e.g., sensing stimuli, responding to stimuli, developing conditioned operants, and bringing about extinction by withdrawing reinforcements).

Positions Against Theory

Despite all the foregoing arguments in favor of theory, some writers have taken positions against theory. They have had in mind theories in the natural sciences and the behavioral sciences.

In Natural Science

Some anti-theory arguments in the natural sciences hold that theory operates against the need for thinking flexibly in doing research and imposes constraints on creativity. Thus Paul Feyerabend (1993), an iconoclast and self-labeled "anarchist" in the philosophy of science, saw theory as keeping thinking away from the "accidents and conjunctures and curious juxtapositions of events" (quoted by Feyerabend, 1993, p. 9, from Butterfield, 1966, p. 66) that characterize original thinking. "Science is essentially an anarchic enterprise, and theoretical anarchism is more humanitarian and more likely to encourage progress than its law-and-order alternatives" (p. 9). As Feyerabend (1993) put it,

> For example, we may use hypotheses that contradict well-confirmed theories or well-established experimental results or both. We may advance science by proceeding counter-inductively ... Hypotheses contradicting well-confirmed theories give us evidence that cannot be obtained in any other way. Proliferation of theories is beneficial for science, while uniformity impairs its critical power. (p. 5)

Here Feyerabend is, however, arguing not against theory but against premature "uniformity" in holding theories. His argument calls to mind philosophy, a discipline that may be said to deal with nothing but theory. Yet philosophy does not stagnate. In support of his antitheoretical position, Feyerabend quoted Lenin, who wrote that "History as a whole is always more varied ... than is imagined by even the best parties" and asserted that its lessons should apply to scientists and method-ologists as well as "parties and revolutionary vanguards" (1993, p. 9). He also noted

Hegel's writing that "What history teaches us is this, that nations and governments have never learned anything from history." And he quoted Albert Einstein (1951) to the effect that

> The external conditions which are set for [the scientist] by the facts of experience do not permit him to let himself be too much restricted, in the construction of his conceptual world, by the adherence to an epistemological system. He, therefore, must appear to the systematic epistemologist as a type of unscrupulous opportunist.

Thus Feyerabend argued that science proceeds according to no rational method. In his view, scientists, in making progress, have violated every logical principle in the philosophy of science literature.

In Literary Studies

Although the field of literary studies is not a science, either natural or social, it is noteworthy that controversies about theory's desirability, similar in tone to those in the natural and social sciences, have arisen in that field. A volume edited by Mitchell (1985) contained an opening chapter titled "Against Theory," which was followed by twelve chapters that the editor considered to be "A Defense of Theory."

In Educational Research

In any case, the differences between the natural sciences and the social and behavioral sciences might be considered to diminish the force of arguments, such as Feyerabend's, based on what has happened in the natural sciences. But the arguments of a sociologist of education (Thomas, 1997) against the desirability of theory in educational research, although not addressed specifically to theory of teaching and not intuitively plausible, deserve attention. They fly in the face of long-accepted doctrine about the uses and consequences of theory. First Thomas notes the ambiguity of the term "theory" – its "multiplicity of meanings" (p. 75). It is highly regarded because of its success "in other fields [such as the natural sciences]" that have "no congruence in education" (p. 76). He argues that

> theory of any kind is thus a force for conservatism, for stabilizing through the circumscription of thought within a hermetic set of rules, procedures, and methods.... [Theory is] an instrument for reinforcing an existing set of practices and methods in education.[It] circumscribes methods of thinking about educational problems and it inhibits creativity among researchers, policy makers, and teachers. (pp. 76–77)

Rajagapolan (1998) replied by holding that "in arguing against theory, Thomas himself ... ends up having to rely on certain well-entrenched theoretical orientations, thus providing an excellent example of the ubiquity of theory in everything we humans do" (p. 337). Also, he wrote, "Underlying Thomas's complaint [about the ambiguity, well-nigh meaninglessness of the word "theory"] is the theory that "every signifier must be attached to one and only one signified" (p. 343). Thomas (1999) then replied, first, that Rajagopalan's view of theory as applying to all "structured thought" was "confusing." Second, he regards such usage of "theory" as "pretentious," "pompous," and having "camouflaging effects."

Thomas uses Piagetian theory as his example of the way in which theories make thinking rigid. Here he somehow overlooked the vigorous revisions of Piaget's theory by many writers, such as Case (1985). The efforts to falsify theories have necessarily entailed original, unshackled thinking. Hellman (1998) described ten of the "great feuds in science," such as those between Pope Urban VIII and Galileo, Wallis and Hobbes, Newton and Leibniz, and, more recently, Derek Freeman and Margaret Mead. His ten accounts, only a small sample of controversies about theory, illustrate the potency of theory in stimulating efforts at refutation. Those efforts have produced genuine advances in human understanding in fields as disparate as gravitation, electromagnetism, evolution, and anthropology.

Knowledge Outcome and Knowledge Use

It is important, in considering these arguments against the desirability of theory, to distinguish between knowledge *outcome* and knowledge *use*.

For Knowledge Producers

It is not true that, in the history of the natural and social sciences, theory has inhibited creativity, originality, flexibility, and intellectual freedom. These assertions face the factual contradiction that, in the natural and social sciences, as the history of major and minor theories shows, and as Popper (1965) contended, theories have indeed provoked indefatigable and creative efforts to falsify them. And theories have survived only when those efforts have failed.

As a precursor of research, theory may serve knowledge production as a stimulus of (a) researchable questions, (b) questions derived from previous thought, (c) hypotheses whose testing increases the likelihood of findings that make sense in relation to prior thought and investigation, and (d) hypotheses that make research findings less in need of an impossible number of confirming replications. So theory is not an indispensable precursor of outcomes of research, although it can be a valuable guide to research that will have scientific payoff.

For Knowledge Users

For knowledge users, theory has the inestimable value of making events or processes understandable. Theory can crystallize otherwise extremely unwieldy stores of factual knowledge into applicable forms. Without theory, we would have, at best, only empirical generalizations that would remain unexplained, isolated from other phenomena, and unrelated to a covering law into which they can be fitted. Theory ties together in an explanatory framework the infinitude of particular instances of the operation of the theory and thus makes them make sense. Theory, in this way, serves knowledge use.

As was easily predictable from where this discussion is located – in a work devoted to the presentation of a theory of teaching – we reject Feyerabend's and Thomas's arguments against theory. To give up the quest for theory means to give up the goal of understanding why the students of some teachers achieve the objectives of teaching at a higher level than the students of other teachers, even when non-teaching factors affecting student achievement – such as student intelligence, home background, and community economic status – are controlled. These differences in student achievement are the cause of much concern, worry, soul-searching, and policy-examining on the part of the millions of parents of the low-achieving students. To abandon the search for relevant theory is tantamount to abandoning the hope of understanding teaching and of improving teaching for those students who most need the improvement.

The arguments against theory will win some acceptance until a valid theory of teaching is developed. In the meantime, the effort should continue. The present attempt may advance education either by surviving attempts at refutation or by stimulating the development of better theory that survives in its turn further attempts at refutation.

Chapter 3
The Evolution of a Paradigm for the Study of Teaching

A paradigm provides a powerful tool for describing a way of viewing a phenomenon, a research program, a set of concepts and variables among which relationships will be sought. A "scientific revolution," in Kuhn's view (1962), occurs when a field of investigation manifests crises, insolvable problems, or anomalies, from which scientists can extricate themselves only by a "paradigm shift" – the adoption of a fruitful new paradigm. As instances of such "paradigm shifts" in the natural sciences, one could name the revolutions in the physical sciences brought about by Newton, Kepler, Lavoisier, Curie, and Einstein. In biology, it was Darwin, Mendel, and Pasteur who led revolutions.

Robert Merton (1955) made one of the earliest uses of the term *paradigm,* in his *Paradigm for the Study of the Sociology of Knowledge.* But it was only after the appearance of Thomas Kuhn's famous monograph, *The Structure of Scientific Revolutions* (1962), that the concept become widely known and valuable in the history and philosophy of science. Since then, the term "paradigm" has been widely used, even in everyday discourse, as a term that stands for "model," "pattern," or "schema." A fixed definition has been elusive, however, as Masterman (1970) showed by identifying 22 different usages of "paradigm" in Kuhn's *Structure.*

The first use of the term *paradigm* in relation to research on teaching occurred in the chapter, "Paradigms for Research on Teaching" (Gage, 1963), written in the spring of 1962 for the first *Handbook of Research on Teaching*, published in February 1963. I had learned the term in 1937 from my undergraduate mentor, B. F. Skinner, when he told me that his *operant* conditioning represented a different paradigm from the *reflex* conditioning of I. P. Pavlov.

The paradigm proposed here fits into the scientific approach (Phillips & Burbules, 2000) to acquiring knowledge and understanding of social and behavioral phenomena, such as teaching. Other, radically different kinds of paradigms have been used for the same general purpose. One of these paradigms is *critical theory* (for example, Gore, 1997; Popkewitz, 1984), which is concerned with the role of teaching and education in developing the *power* of social class, gender, and ethnic segments of society.

This chapter's paradigm for the study of teaching evolved during the twentieth century. It consists of six basic *categories* of related concepts that underlie the

N.L. Gage, *A Conception of Teaching,*
DOI: 10.1007/978-0-387-09446-5_3, © Springer Science+Business Media, LLC 2009

theory presented here. Variables are concepts whose referents can take on forms or values that differ from one another, qualitatively or quantitatively. For example, gender is a concept that varies "qualitatively." Height is a concept that varies "quantitatively." In this chapter the variables will be described and illustrated in terms of how they have entered into logical and empirical studies of teaching.

Evolution of the Paradigm

In research on teaching, the paradigm has evolved from the cumulative contributions of students of teaching in the period from the 1890s to the 1980s. The evolution began with a paradigm implicit in the empirical research design used by Joseph Mayer Rice (1897; reprinted in Rice, 1913). The subsequent insertion and modification of additional categories took place as successive thinkers filled in the overall conception of what needed to be considered in a full attempt to explain teaching.

The Process-Achievement Paradigm

Rice (1897; see also Banerji, 1988) studied the relationship between variables in two categories that would later be called *process* and *achievement*. In his study the *process* variable was the amount of time teachers devoted to teaching spelling. The *achievement* variable was the average score of the teacher's students on a spelling test. The study of this relationship was later termed the "process-product paradigm," shown in Fig. 3.1a, whose two boxes contain the two categories of variables, and the arrow represents the relationship between them.

The Criterion-of-Teacher-Effectiveness Paradigm

Later, unaware of Rice's work, I identified the similar "criterion-of-teacher-effectiveness" paradigm, shown in Fig. 3.1b (Gage, 1963, p. 114). It differed from Rice's process-product paradigm in allowing for mere *correlates,* such as promising predictors of teaching effectiveness in the form of teacher characteristics, in addition to the teaching *process* variables, such as Rice's, that connoted *determiners*

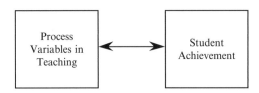

Fig. 3.1a The process-achievement paradigm (Rice, 1897, in Rice, 1907).

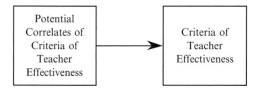

Fig. 3.1b The criterion of teacher-effectiveness paradigm (Gage, 1963, p. 114).

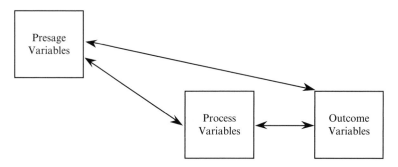

Fig. 3.2 The paradigm resulting from the insertion of presage variables.

of the achievement variables. Also, it admitted criteria of teaching other than student achievement – criteria such as principals' ratings of teachers and teachers' length of experience in teaching.

Presage Variables

Mitzel (1960) introduced the concept of *presage* variables. The term denotes dimensions of teacher personality and teachers' experience in teacher education programs that are considered to be potential predictors, or "presages," of teaching effectiveness (Fig. 3.2). As Mitzel (1960) described them,

> presage variables, so-called here because of their origin in guessed-at-predictive-value, are from a logical standpoint completely removed from the goals of education. Their relevance depends on an *assumed* or conjectured relationship to other criteria, either process or product... There are at least four types of presage variables... (a) teacher personality attributes, (b) characteristics of teachers in training, (c) teacher knowledge and achievement, and (d) in-service teacher status characteristics. (p. 1484)

Context Variables

Mitzel (1957, cited in Gage, 1963, p. 121) also introduced the category consisting of *context* variables, or, as Mitzel named them, "contingency factors." These describe the setting, or environment, in which the teaching goes on: the relevant characteristics of the culture in the nation, region, community, school, classroom, family, and student body.

Adding this category to the previous three resulted in the four-category paradigm used by Biddle (1964) with different terminology. For the presage category, Biddle's term was "teachers' properties and formative experiences." He called the context category "school and community contexts" and "classroom situations." For the process category, his terms were "teacher behaviors" and "immediate effects on pupil responses." For the achievement category, his term was "long-term consequences," including pupil achievement and adjustment.

But Dunkin and Biddle (1974), in their comprehensive, insightful review of research on teaching, adopted Mitzel's terminology: presage, context, process, and achievement (see Fig. 3.3). More recently, the Center for Research on the Context of Teaching (McLaughlin & Talbert, 2001), described one of its monographs as dealing with teachers' professional communities in American high schools at the end of the twentieth century... We describe how the work of teaching differed – in classroom practice, in colleague relations, and in experienced careers – across three different types of communities we found in high schools and subject departments ... (p. 2).

The Teacher's-Thought-Processes Category

B. O. Smith (1961, p. 92) contributed the category for teachers' thought processes when he called attention, in his "pedagogical model," to what he called the teacher's "intervening variables," namely, the teacher's thought processes, which followed

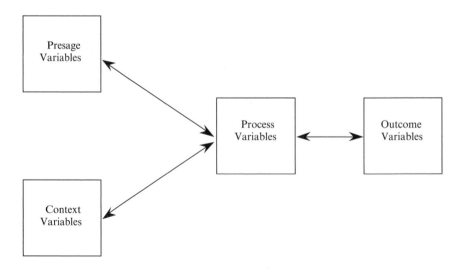

Fig. 3.3 The paradigm resulting from the addition of context variables (Mitzel, 1957).

upon the presage and context variables and were antecedents of the process variables. Shulman (1975) and his fellow members of a planning conference panel on "Teaching as Clinical Information Processing" at the National Institute of Education's National Conference on Studies in Teaching (Gage & Viehover, Eds., 1975), brought this category to the fore by elaborating on its potential theoretical significance:

> Thus an understanding of how teachers cognitively construct the reality of teaching and learning remains central to the achievement of NIE's overall goal of developing the means to improve the provision, maintenance, and utilization of high quality teaching personnel. A teacher may possess the full range of relevant instructional skills, but if he is unable to diagnose situations in which a particular set of those skills is needed, the skills alone will be insufficient. (Shulman et al., 1975, p. 2)

A decade later, Shulman (1986a) elaborated concern with teacher *knowledge* as an important aspect of the teacher's thought processes. Here he saw a "missing program" in that researchers had neglected the *content* of teaching. He distinguished between three kinds of content knowledge: "subject matter knowledge, pedagogical knowledge, and curricular knowledge":

> *Subject matter knowledge* is that comprehension of the subject appropriate to a content specialist in the domain... *Pedagogical knowledge* refers to the understanding of how particular topics, principles, strategies, and the like, in specific subject areas are comprehended or typically misconstrued, are learned and likely to be forgotten. *Curricular knowledge* is familiarity with the ways in which knowledge is organized and packaged for instruction in texts, programs, media, workbooks, other forms of practice, and the like. (p. 26)

Subsequently, Clark and Peterson (1986) were able to review a considerable body of research literature on teachers' thought processes (Fig. 3.4).

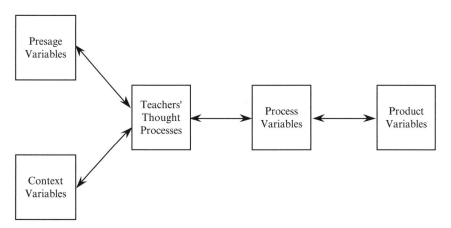

Fig. 3.4 The paradigm as modified by the addition of Teacher's Thought Processes (B. O. Smith, 1961; Shulman et al., 1975) and the Content of Teaching (Shulman, 1986a).

This category consists of the actions and interactions of teachers and students in the classroom (process) and the subject-matter taught (content). Content was classified by *A Taxonomy for Learning, Teaching, and Assessing* (Anderson et al., 2001). It distinguished between four types of knowledge to be learned: factual knowledge, conceptual knowledge, procedural knowledge, and metacognitive knowledge. (These types will be described below.)

The *Taxonomy* also distinguished between six categories of cognitive process: remember, understand, apply, analyze, evaluate, and create. Because of our emphases, described and justified in Chap. 8, we spell out only the two subcategories of *remember*: recognize and recall, and only one of the six *sub*categories of cognitive process: *understand*.

The Student's Thought Processes Category

B. O. Smith (1961) also included in his model the students' thought processes. He considered these processes to follow classroom process variables and precede achievement variables. When students' thought processes are considered as stable characteristics, such variables are considered context variables – part of the situation in which the teaching occurs.

When such processes occur momentarily during classroom discourse, they are considered "student's thought processes." Winne (1982, 1987, 1995), Winne and Marx (1983), and Marx and Winne (1987) reviewed pertinent research and developed a detailed rationale for their concern with the thought processes of students (Fig. 3.5). They wrote:

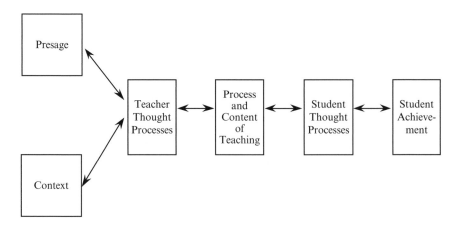

Fig. 3.5 The paradigm as modified by the addition of student's thought processes (B. O. Smith et al., 1967; Doyle, 1977; Marx & Winne, 1987).

Put briefly, between the teacher behavior and students' accomplishments assessed after teaching, we inserted an explicit place for students' cognitions to occur... The cognitive mediational model [so termed by Marx and Winne,1987] also recast the locus of causation in the performance-based model. Instead of teacher behaviors serving as the causes of students' learning, teacher behaviors in the cognitive mediation model became signals for the students to use certain cognitions to learn content. The students' cognitions were considered the causes for learning. (Marx & Winne, 1987, pp. 270–271)

The long-overdue realization that *what* the teacher teaches deserves as much attention in research on teaching as *how* she teaches should lead to a major revision in the approach to research on teaching. Also, the terms "process–product" or "process–outcome," with their factory-like connotations, should be dropped. The term "process↔content–achievement" should replace them. (The bidirectional arrow is intended to symbolize the interaction between process and content.) That is, the content of teaching should no longer be omitted from the term that describes the research concerned with relationships between teaching and student achievement. Accordingly, it follows that process↔content–achievement research should henceforth be the term for research aimed at discovering relationships between teaching – how and what teachers teach – and what students learn.

As Fig. 3.6 shows, the six categories are labeled with Capital letters. The categories can be categorized into two sets: (a) those – Categories A, B, C – that are logically *antecedent* of Category D: and (b) Categories E and F that are logically *subsequent* to Category D.

The relationships between all 15 pairs of categories that are either logically prior or logically subsequent to the process↔content events of teaching are represented by the two-way arrows numbered 1–15 connecting pairs of categories. For example, student achievement can affect *subsequent* student's thought processes, and the *prior* process↔content of teaching can influence *subsequent* teacher's thought processes.

The Variables in the Categories

Here we examine and give examples of the variety of concepts, or variables, that fall within each of the categories.

The Presage Category

This category consists of such characteristics of the teacher as gender, age, and years of experience. It also comprises *traits* (stable characteristics, such as intelligence, knowledge about ways of teaching, and introversion-extraversion) both in general and in the context of a specific subject matter. The teacher's cognitive abilities affect her grasp of the subject matter in all its variations and complexities, including the previously identified pedagogical content knowledge introduced by Shulman (1986b).

The teacher's knowledge of the subject she is teaching – her content knowledge – affects the way she presents, explains, illustrates, and demonstrates the content she wants her students to learn. The teacher's experiences in a teacher-education program may influence her conceptions of teaching and her implicit values about how teaching should proceed. Her previous experience as a teacher influences her security and optimism about teaching.

The presage category also includes the teacher's stable affective characteristics: intentions, beliefs, attitudes, values, appreciations, and the like, as traits that the teacher has acquired from experience, including experience in a teacher education program.

Her *personality* may affect the general demeanor of her engagement with her tasks and students. Currently, the so-called "Big Five," dimensions of personality, widely accepted by psychologists (see, for example, Cutchin, 1999), illustrate one way in which personality is given dimensions. Its five dimensions are the degree to which a teacher tends to be (a) hostile *versus* agreeable, (b) introverted *versus* extroverted, (c) impulsive *versus* conscientious, (d) neurotic *versus* emotionally stable, and (e) intellectually narrow *versus* intellectually open.

The Context Category

This category consists of characteristics of the nation, region, community, school, and class in which teaching takes place. The *community* characteristics include urban or rural; average income; ethnicities. The *school* characteristics include student-body size, resources for teaching, the student-teacher ratio. The *class* characteristics include the students' socioeconomic backgrounds, cognitive abilities, ethnicities, previous school achievement, and the students' heterogeneity along these dimensions.

The Teacher's Thought Processes Category

This category comprises the *momentary* thought processes that deal with cognitive aspects of her teaching – such as the content being taught, its organization, its facts, concepts, and principles. Her thought processes also deal with affective aspects of her teaching: her attitudes, motivations, and values, and emotionally loaded behaviors. Her thought processes deal especially with her momentary use of pedagogical content knowledge.

When the teacher's thought processes are a *stable* characteristic of the teacher, the teacher's thoughts become presage variables – how her ideas about process and content interact to affect her teaching. When they occur in the midst of the process of teaching, they are transitory states. Both stable and transitory thought processes occur before, during, and after the teacher interacts with her students. Jackson (1968) identified two kinds of teacher thought processes: preactive and interactive. *Preactive* thought processes, such as planning, occur before she interacts with her students. They call upon her knowledge, beliefs, and values about teaching, learning, and the curriculum. The successes and failures of recent interactions affect the teacher thought processes in planning for the next round of teaching.

Interactive thought processes, occur during her interactions with students, which Jackson found to occur as often as a thousand times per day. She thinks, for example, about the explanation she is giving, the questions she asks, her students' responses, and her reactions to students' responses. She also thinks about the success of a recent exchange with students, about her students' comprehension, and about the next few steps in her teaching.

Post-interactive thought processes are those she engages in after she has had a class period with students. These thoughts deal with such things as her satisfaction with the way the lesson has gone; whether she needs to change her approach; her perception of students' interest, attention, motivation, and comprehension. Clark and Peterson (1986) reviewed research on teachers' planning, interactive thought processes, and post-interactive thought processes.

The Variables in the Process ↔ *Content of Teaching Category*

As noted earlier, only the verbal (and not the nonverbal) aspects of process will be given detailed attention. Note that we have distinguished between two types of process: (a) *cognitive* processes, which refer to the student's mental activities while learning and (b) *teaching* processes, which refer to what the teacher is doing: verbal behavior; cognitive or social-emotional interactions with students; interactions with the whole class, subgroups of the class, or individual students. Of these we give special attention to the types of verbal interaction.

The verbal process part. The verbal aspects of the process of teaching will be those identified by Bellack, Kliebard, Hyman, & F. L. Smith (1966), Hoetker and Ahlbrand (1969), and, with different terminologies, by Mehan (1979) and Goodlad (1984):

(a) *structuring* (typically by a teacher, setting forth and organizing the content),
(b) *soliciting* (typically by a teacher asking questions of students),
(c) *responding* (typically by a student answering a teacher's question), and
(d) *reacting* (typically by a teacher after a student's response).

The content part. The subject-matter will be as specific as the teacher's purposes require. Examples of possible subject-matter-specific process variables are those used in teaching paragraph analysis in reading, in teaching the use of the semicolon in writing, in teaching the addition of numbers in arithmetic, in teaching the gas laws in science, in teaching Shakespeare's metaphors, and in teaching the checks and balances in the U.S. Constitution.

Berliner (1989) argued for the study of teacher's thought processes in relation to student achievement. The question to be answered was, Do contrasting groups of teachers, that is, high and low groups of teachers on any teacher-thought-process variable, have different effects on student achievement by virtue of their connection with kinds of teacher behavior? Research on the effects of teacher thinking could study the thought processes of groups of teachers selected for their differing degrees of effectiveness in fostering student achievement. Such research would yield knowledge of Type 12 connections, that is, relationships between teacher's thought processes and student achievement. In the process, the research could also throw light on relationships of Type 10 (between teacher's thought processes and process ↔ content of teaching) and Type 11 (between teacher's thought processes and student's thought processes).

The Student's Thought Processes Category

This category includes

> student perceptions, expectations, attentional processes, motivations, attributions, memories, generations, understandings, beliefs, attitudes, learning strategies, and metacognitive [monitoring-own-thoughts] processes that mediate achievement. (Wittrock, 1986b, p. 297)

All these take the form of momentary states, not long-lasting traits.

The Student Achievement Category

This category represents the goal of all the foregoing categories. It includes *achievement* of *cognitive* objectives and can also refer to achievement of *social-emotional* objectives (e.g., adjustment and attitude) and *psychomotor* objectives (e.g., gymnastics and dancing). As was noted in Chap. 2, the present theory deals only with cognitive achievement.

At the end of some teaching, do the students *know*, in the sense of being able to recall or recognize, what they should know? Do they *understand* in the sense of being able to summarize, explain, translate, and apply what they should be able to? Beyond the cognitive objectives of a specific lesson, or course of study, educators also value *metacognitive skills*: students' abilities to monitor and control their own thought processes for the purpose of fostering their own achievement.

This category contains the evidence of the student's achievement of the objectives at which the teaching was aimed. *The Taxonomy for Learning, Teaching, and Assessing* (Anderson, Krathwohl et al., 2001) describes student achievement along two dimensions:

(a) *the Knowledge Dimension* with four segments: Factual Knowledge, Conceptual Knowledge, Procedural Knowledge, and Metacognitive Knowledge, and

(b) *the Cognitive Process Dimension* with six segments: Remember, Understand, Apply, Analyze, Evaluate, and Create.

Thus, a cognitive objective of teaching – a desired outcome – consists of a pairing of a kind of *Knowledge* (facts, concepts, procedures, and metacognitions) with a kind of *Cognitive Process* (remember, understand, apply, analyze, evaluate, and create). For example, a cognitive educational objective might consist of (a) remembering (cognitive process) the multiplication table (type of factual knowledge); or (b) understanding (cognitive process) the function of analogy in a passage of prose (conceptual knowledge).

The Change from "Process" to "Process ↔ Content"

What came to be known as process-product, or process-outcome, research on teaching focused on how the process of teaching relates to student achievement. Such research was prominent in research on teaching from the 1960s to the 1990s. It is represented in Fig. 3.6 by the relationships of Type 14 between variables in Category D and variables in Category F.

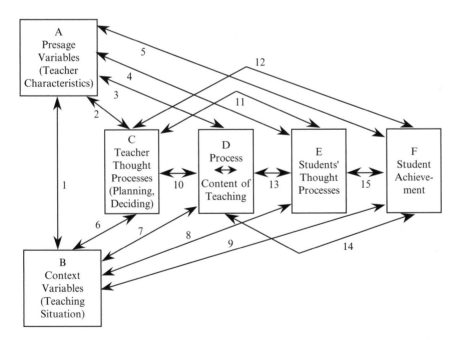

Fig. 3.6 The paradigm with all six categories, lettered A-F, and all 15 two-way arrows, numbered 1–15, indicating the two-way influences between all possible pairs of categories.

The revised version of this category includes the Content of Teaching along with the Process of Teaching so as to result in a category labeled (Process ↔ Content), which refers to both the teaching process and the content being taught. *But "pedagogical content knowledge" suggests that it is merely the content of the teacher's mind. Here, however, content is part of the teacher's action and behaviors. We hypothesize that this category will improve the explanatory power, the predictive power, and the improvability of student achievement beyond that produced by process-product (without content) research.*

On the Process Side

Here is *how* the teacher teaches. Those actions and interactions take such forms as the recitation and the discussion group. Here the variables take such forms as the comprehensibility of explanations, the cognitive level of the questions the teacher asks (ranging from remembering,... to creating), wait-time (the number of seconds the teacher waits after asking a question (see Rowe, 1974), and the helpfulness of the teacher's reaction to the student's response to a question. On the students' side, these variables include the amount of *academic learning time*, or the time during which a student attends to learning tasks that permit high success rates and are relevant to academic objectives (Berliner, 1990). Affective components of the teacher's actions and interactions take such forms as the teachers' warmth, supportiveness, permissiveness, and authoritarianism.

The classroom processes also comprise the students' interactions with other students and the ways in which they influence one another. Nuthall and Alton-Lee, (1998) described careful studies of students' interactions with other students.

On the Content Side

Here belongs what the teacher teaches – the subject matter: the facts, concepts, procedures, and metacognitions in the form in which they are specified in the curriculum of the subject being taught. Here one major variable is *instructional alignment* (S. A. Cohen, 1987, 1995), or the degree to which the content taught and the content assessed (tested) are congruent with each other and with the objectives of the teaching. The term *assessment* is used here to stand for all kinds of assessment: multiple-choice tests, essay tests, observations of student performance, interviews, portfolios (collections of students' products), norm-referenced tests (tests that compare one student's performance with that of a norm group of students), and criterion-referenced tests (tests that compare onestudent's performance with an established standard).

An example of the potential importance of combining content with process, so as to replace process-product research with (process↔content)-product research can be seen in the "somewhat disappointing" (D. W. Ryan et al., 1989, p. 28) results of the massive process-product investigation of teaching in nine countries. Carried out by the International Association for the Evaluation of Educational Achievement (IEA), the research dwarfed previous efforts to find relationships between teaching and achievement.

In seeking to explain the disappointing results, we note the first of D. W. Ryan's "generalizations": "Within countries, teachers differ greatly in what they teach relative to what is tested" (p. 29). Thus, the study collected data primarily on teachers' processes, and found those processes disappointingly related to student achievement, which was, of course, affected by the content of the teacher's teaching, or the degree to which that content gave her students "opportunity to learn." In Chap. 6, we refer to "instructional alignment," or the degree to which what is taught is similar to what is tested. The teacher's teaching should be described not only in terms of her process but also in terms of her content.

A variable related to instructional alignment is *transfer demand* (S. A. Cohen, 1987), or "the degree to which the stimulus conditions of instruction match the stimulus conditions of assessment" of achievement. This variable is considered to be inversely related to instructional alignment; the higher the alignment, the lower the transfer demand.

Another major content variable, further considered in Chap. 8, is *cognitive load* (Sweller, 1999), or the degree to which the (process↔content) of teaching (a) requires the students to split their attention, that is, pay attention to two or more sources of stimuli at the same time, such as a visual presentation and a not-closely-integrated oral presentation, (b) presents tasks with interactive elements, or concepts and ideas that cannot be understood one at a time because the meaning of one depends on the meaning of one or more others; or (c) requires conventional problem solving as against letting students study worked-out examples.

Similarly, a strong case can be made for concern with student's thought processes in attempts to explain, predict, or improve student achievement. Winne and Marx (1987), in particular, argued for the value of incorporating data on student's thought processes (Category E) in the search for improved explanations of teaching's effects on student achievement.

The teacher's knowledge of the content being taught matures as the teacher acquires experience. As Gage (1979) put it,

> Experienced teachers have often noted that their years of teaching have given them extensive repertoires of effective explanations, demonstrations, illustrations, examples, diagrams, and anecdotes for the myriads of concepts and principles that they teach and the many understandings and skills that they help their students acquire. Just as medicine and engineering have not shrunk from these levels of complexity in their research and development, so research on teaching may also need to do the fine-grained work that will yield better ways of teaching a specific skill (such as long division) to a specific kind of pupil (such as an anxious fifth-grader). (p. 273)

Shulman (1986a) formulated the concept of *pedagogical content knowledge* to integrate the process and content of teaching: He wrote that

Mere content knowledge is likely to be as useless pedagogically as content-free skill. But to blend properly the two aspects of the teacher's capacities requires that we pay as much attention to the content aspects of teaching as we have recently devoted to the elements of the teaching process. (p. 8)

Pedagogical content knowledge manifests itself in the degree to which the teaching of a particular segment of content reflects such kinds of knowledge as

(a) the main, or most frequently taught, concepts in the content area,
(b) different conceptions of the subject, such as how it has been changing over time, what topics have been recently become prominent, and which ones are on the way out;
(c) metaphors, similar to similes, without the "like," for explaining a concept or principle;
(d) analogies that help clarify a particular concept or principle by likening it to a more familiar or better understood concept or principle;
(e) examples, which give concrete and familiar instances of a particular concept or principle;
(f) mnemonics, or memory aids, for remembering such things as the rhyming scheme of a sonnet (abab, cdcd, efef, gg); and
(g) the difficulty of various topics, or the degree to which students find them hard to understand and the reasons for which they are difficult.

Gess-Newsome and Lederman (1999) provided applications of pedagogical content knowledge to science education.

Relationships between All Possible Pairs

On the pedagogical-content-knowledge side (Shulman, 1986b, 1987), the teacher applies what she knows about pedagogy to the particular details of the topic within the content being taught. Here belong the types of clarification (analogies, illustrative cases, acronyms, structures, etc.) that teachers acquire through their experience in teaching a given subject matter.

A Paradigm for the Study of Teaching

So the present conception of *teaching* is that it becomes the proper subject of theory development when it is specified for the purpose of fostering a certain kind of *learning* appropriate to a certain kind of *curriculum*. Figure 3.6 brings together the resulting six categories of variables. The two-way arrows connecting pairs of categories denote relationships between the concepts in each pair of categories – relationships of the kind that can be revealed by case studies, correlational studies, or experiments.

The fifteen possible relationships shown by the arrows in Fig. 3.6 need to be evaluated as to their promise for theory and research. It is conceivable that strong theoretical and empirical relationships could be found in all fifteen cases.

Two-way Relationships Between Pairs of Categories

The influence between a pair of categories can go in either direction, depending on which variable in the two categories occurs first. The numbers in the following list are the same as those of the 15 two-way arrows shown in Fig. 3.6.

The following pairs of examples illustrates the relationship between each pair of categories, first in one direction, then in the other. The symbol $><$ stands for two-way influences in which the $>$ signifies influence from the first-named member of the pair, and the $<$ indicates the opposite direction of influence. All in all the examples illustrate the rich variety and complexity of the phenomena inherent in teaching.

1. *Presage $><$ Context.* A teacher with a need for simplicity may tend to create a certain student grouping for her teaching. *Or* a school district with caring parents may be willing to employ teachers with varied backgrounds.
2. *Presage $><$ Teachers' Thought Processes.* Teachers with high scholastic abilities make more appropriate plans for classroom processes. *Or* teachers' positive feelings about teaching lead to longer careers as teachers.
3. *Presage $><$ Process↔Content of Teaching.* Teachers with higher verbal aptitude hesitate in offering alternative explanations when students do not comprehend the first explanation. *Or* confusion in the teacher's conduct of mathematics lessons may induce teachers to improve their competence in mathematics.
4. *Presage $><$ Student Thought Processes.* Teachers with a high level of competence in mathematics may influence students to see the logic of mathematical ideas more clearly. *Or* students' puzzlement may influence teachers to improve their pedagogical content knowledge.
5. *Presage $><$ Student Achievement.* Teachers with rural backgrounds may foster higher student achievement in botany. *Or* low student achievement in mathematics may affect teacher motivation for teaching science.
6. *Context $><$ Teachers' Thought Processes.* Poor laboratory equipment for high school chemistry may motivate teachers to think of substitutes for missing equipment. *Or* teachers' eagerness to teach *Macbeth* may lead them to develop an Elizabethan theatre in the school's auditorium.
7. *Context $><$ Process↔Content of Teaching.* The arrival of a new computer factory in town may enable teachers to raise the level of science content taught in the community's schools. *Or* the English teacher's way of teaching poetry may make the town library acquire more copies of Shakespeare's sonnets.
8. *Context $><$ Student's Thought Processes.* Smaller class size may make students less hesitant about asking for help. *Or* student enthusiasm about developing a school newspaper may lead to the school's purchase of a small printing press.

9. *Context >< Student Achievement.* Reducing class size seemed to improve student achievement. *Or* low student achievement in writing made the school board willing to improve school libraries.

10. *Teachers' Thought Processes >< Process↔Content of Teaching.* A teacher who pondered about how to improve her classroom management skills was then able to keep her class more engaged in learning activities. *Or* the teacher's success in fostering a genuine discussion made her think about why it had gone so well.

11. *Teachers' Thought Processes >< Student's Thought Processes.* A teacher's high expectations of her students' performance may raise the students' aspirations. *Or* students' formulations of the content may influence their teacher's perceptions of her students.

12. *Teachers' Thought Processes >< Student Achievement.* Teachers' beliefs about the importance of certain scientific facts may influence student comprehension of that content. *Or* student achievement of high-level cognitive objectives may make the teachers value that content more.

13. *The Process↔Content of Teaching >< Student's Thought Processes.* Teachers' interactions with students may influence their students' aspirations. *Or* students' interest in the content may make teachers use the same approach next time.

14. *The Process↔Content of Teaching >< Student Achievement.* A debate between the two groups of students resulted in improved student understanding of the issue. *Or* students'inability to apply a principle influenced the teacher to arrange for an exhibit of illustrative applications.

15. *Student's Thought Processes >< Student Achievement.* Students' pessimism affected their learning to multiply fractions. *Or* student achievement of the ability to apply a principle raised their motivation to learn the subject matter.

Ways of Describing the Process of Teaching

Suppose the process of teaching is described on the basis of the positions of any given set of processes on any set of dimensions descriptive of any process. One such dimension is teacher-centered versus student-centered.

Similarly, it might be possible to describe student achievement of objectives on the set of dimensions. One such set of dimensions is presented in the *Taxonomy for Learning, Teaching, and Assessing* (Anderson & Krathwohl, Eds., 2001), which makes possible the description of educational achievement in any subject matter on the basis of a set of six dimensions. Ultimately, the theory might aim at predicting connections between the profile of the teaching description and the profile of the achievement description.

The fifteen possible relationships shown by the arrows in Fig. 3.6 need to be evaluated as to their promise for theory and research. Although it is conceivable that strong theoretical and empirical relationships could be found in all fifteen cases, the present treatment of them focuses only on those that proceed from left to right to the very next category of the paradigm. This means that relationships will be considered in detail only between the following sets of categories:

1. Categories A and B in relation to Category C
2. Category C in relation to Category D
3. Category D in relation to Category E
4. Category E in relati on to Category F

1. The relationships of Categories A (presage variables) and B (context variables) to Category C (teacher's thought process): Psychologists have often endorsed the proposition that behavior is a function of the personality interacting with the environment.

One basis for developing theory would be to classify teaching methods and practices and make profiles of them on the basis of specified dimensions. These could then be used as descriptions of the classroom processes used by teachers.

How do the three components of the ternary relationship – teaching, learning, and curriculum – fit into this paradigm? The answer is that the conceptual antecedents of teaching are the presage, context, and teacher's thought-process categories. That is, all of the variables in these categories shape the content and processes of teaching. And the classroom processes, in turn, influence the students' thought processes and achievement.

The theory of teaching should particularly explain the relationships between classroom process↔content and student achievement. The relationships between presage variables, context variables, teacher's thought processes and the process↔content of teaching have been addressed by Schoenfeld (1998). In his *Toward a Theory of Teaching-in-Context*, he attempted to "provide a detailed theoretical account of how and why teachers do what they do 'online,' – that is, while they are engaged in the art of teaching" (p. 1). He was concerned with "the ways in which the teachers' goals, beliefs, and knowledge interact, resulting in the teachers' moment-to-moment decision-making and actions" (p. 1). He was engaged in "explaining, at a fine-grained level of detail, how and why teachers make specific decisions and take specific actions as they are engaged in teaching" (p. 1).

Other categories of the paradigm – presage, context, and teacher's thought process – are central to Schoenfeld's focus on *teachers' specific decisions and specific actions as the dependent variable*. Connection between these categories and the heart of teaching, what goes on in the classroom, is extremely consequential. The present focus is on teachers' decisions and actions as the independent variables with *student achievement as the dependent variable.* Thus, the process↔content-achievement paradigm will be the focus of the present theory of teaching.

The category that intervenes between classroom process↔content and student achievement consists of student's thought processes. Although Student's Thought Processes are neither visible nor audible and, hence, are not directly observable, they can be studied by inferences from other kinds of data described in Chap. 8. Depending on the specific forms of each of these processes, they can either foster or hamper learning.

Intra-Category Relationships

Each category contains many concepts. Relationships between concepts can be intra-category, that is, relationships between concepts within a single category, or inter-category, relationships between concepts in different categories.

Intra-category relationships elucidate the structure of the category – its dimensions and divisions. Thus, within the presage category we could seek relationships between the teacher's scholastic aptitude and her pedagogical content knowledge, or between her socioeconomic background and her sensitivity to social-class differences among students.

Inter-Category Relationships

Inter-category relationships show how a concept or variable in one category relates to one in a different category. One frequently studied relationship of this kind is the process↔content-achievement relationship. Perhaps the most frequently studied inter-category relationship has been that between the "process," a part of the process↔content category, and the "outcome," as the student-achievement category was called. In 1986, it was characterized as "easily the most vigorous and productive of the programs of research on teaching during the past decade" (Shulman, 1986a, p. 9). The reason for this strong interest in process↔content-outcome research is easily understood. The *raison d'être* of teaching is beneficial effects on students. However interesting relationships between other pairs of categories may be, most persons concerned with teaching are interested in understanding and improving its effectiveness. In the present terminology that interest takes the form of process↔content-achievement research.

Multivariate Relationships

Relationships can also be multivariate in the sense that two or more variables can correlate with a third, as in a multiple correlation. An example would be $R_{x.yz}$, the correlation between x (achievement) and the student's (y) scholastic aptitude and (z) prior knowledge. Or the relationship between two variables can be studied in terms of how a third variable is held constant, as in partial correlation, $r_{xy.z}$. An example would be the correlation between x (achievement) and y (scholastic aptitude) with z (socioeconomic status) held constant.

The large number of concepts or variables within each of the six categories indicates the enormous number of relationships that could possibly be studied. This enormousness requires that only a few of these relationships be chosen for theorizing, that is, for an attempt to explain their relationships. Choices here will reflect the investigators' interests and concerns.

Presage variables can be studied in relationship to context variables to determine whether teachers with higher scholastic achievement tend to be employed in wealthier communities. Presage variables can be studied in relation to teacher thought process variables to determine whether differences in say, teacher scholastic aptitudes tend to be associated with differences in teacher's thought processes. Similarly, context variables can be studied in relation to teacher's thought processes to determine whether teachers in smaller classes tend to give more thought to the learning difficulties of individual students. Or a presage variable such as the teacher's experiences in her teacher education program can be studied as to whether they are related to the thoroughness of the teacher's planning of her teaching. Similarly, relationships can be sought between teacher's thought processes and their teaching as described in the process↔content category.

An analogy with medical research would recognize the importance of all of the research that leads to the development of effective forms of diagnosis and treatment. For diagnosis, the development of X-rays, the identification of bacteria and viruses, the development of instruments such as microscopes, magnetic resonance indicators (MRIs), and so on, are undertaken. For treatment, the development of pharmaceuticals and surgical techniques are a part of the vast medical enterprise. The development of effective means of diagnosis and treatment requires the contributions of all of the foregoing research enterprises.

But, in the end, medical science and technology focus on the attainment of desirable results – in bringing about cures, recoveries, good health, and longevity – and on the treatments that have such effects. In medicine, this focus has led to "treatment-process research" (Dubois & Brown, 1988; Wennberg, 1989), an approach sharing many of the attributes of process↔content-achievement research on teaching. Such medical research seeks to identify relationships between the kind of treatment given patients and the patients' subsequent health.

The present theory is aimed at explaining relationships between process↔content variables and student achievement variables. The relationships can take the form of correlations or effect sizes i.e., standardized differences between the effects of treatments.

Content refers to what is taught: the subject, the concepts within that subject, the relationships between concepts in that subject, the facts, and principles in that subject, and so on. Different formulations of content apply to different subject matters. The content of English literature is formulated along lines that differ markedly from the content of chemistry.

Instructional alignment

But one variable characteristic of the content that is hypothesized to correlate with student achievement of the objectives of the teaching in that content is the content's "instructional alignment." This term designates the similarity, or congruence, between the content taught and the assessments of achievement in that content area.

One hypothesizes that achievement will be high. If the instructional alignment is high. That is, there should be a positive correlation between the instructional alignment of the teaching and the student achievement of the objectives of that teaching.

The process component of the process↔content category refers, to put it simply, to the way in which the teaching has gone on. Here, the number of descriptive concepts is also indefinitely large. But the one on which we focus first is instructional time, or the aspects of the duration of time during which the teaching of a particular part of subject matter, or content, has proceeded (Ben-Paretz & Bromme, 1990).

Another concept in the content part of the process↔content category, is the resultant of the teacher's *pedagogical content knowledge* (Shulman, 1986a; 1987; Gess-Newsome & Lederman, Eds., 1999). Such pedagogical content knowledge manifests itself in the degree to which the teaching of a particular segment of content reflects (a) the main, or most frequently taught concepts in the content area, (b) different conceptions of the subject such as how it has been changing over time, what topics have been recently become prominent, and which ones are on the way out; (c) metaphors, similar to similes, without the "like," (d) analogies that helped clarify a particular concept by likening it to a more familiar or better understood concept or principle; (e) examples, which give concrete and familiar instances of a particular concept or principle, (f) mnemonics, or memory aids, for remembering such things as the rhyming scheme of a sonnet (abab, cdcd, efef, gg); and (g) the difficulty of various, i.e., topics or the degree to which students find them hard to understand.

Chapter 4
A Conception of the Process of Teaching

How teaching happens, what the teacher and students say and do, what students experience as they see and hear the teacher and their classmates in the classroom – all these and more add up to the *process of teaching*. "Teaching" encompasses what teachers do in helping their students learn and perform the tasks – listening, thinking, speaking, reading, writing, solving problems, answering questions, investigating, and so on – as prescribed, recommended, or suggested by the teacher. The process of teaching should be integrated with the content of teaching. The facts, ideas, knowledge, understandings, concepts, principles, activities, theories, procedures, and the like, help students understand the curriculum.

Since antiquity, Plato and other philosophers have devised, demonstrated, and advocated various processes of teaching. Broudy (1963) developed the historical narrative on processes of teaching into the nineteenth century.

One example of eighteenth century writing on the process of teaching is Rousseau's *Émile*, which exemplified a "romantic" view of good teaching – one that gave students well-nigh complete freedom to explore on their own. Wallen and Travers (1963) wrote about Rousseau and his predecessor, Froebel, as follows:

> Like Rousseau .... [Froebel was influenced by the concept that development will proceed harmoniously of its own accord if the child is provided with a suitable environment. Emphasis was placed on the individual worth of each child, and teacher behavior had to be such that it did not do violence to the natural laws of the growing organism. The teacher must be permissive so that the natural process of development will not be violated. (p. 455)
>
> Skinner (1974), the famous behaviorist, offered this comment on Rousseau's *Émile*: His name is Émile. He was born in the middle of the eighteenth century in the first flush of the modern concern for personal freedom. His father was Jean-Jacques Rousseau. But he had many foster parents, among them Pestalozzi, Froebel, and Montessori, down to A. S. Neill and Ivan Ilich. He is an ideal student. Full of goodwill toward his teachers and his peers, he needs no discipline. He studies because he is naturally curious. He learns things because they interest him.
>
> Unfortunately, he is imaginary. He was quite explicitly so with Rousseau, who put his own children into an orphanage and preferred to say how he would teach his fictional hero, but the modern version of the free and happy student to be found in books by Paul Goodman, John Holt, Jonathan Kozol, or Charles Silberman is also imaginary. Occasionally a real example seems to turn up. There are teachers who would be successful anywhere – as statesmen, therapists, businessmen, or friends – and there are students

N.L. Gage, *A Conception of Teaching*,
DOI: 10.1007/978-0-387-09446-5_4, © Springer Science+Business Media, LLC 2009

who scarcely need to be taught, and together they sometimes seem to bring Émile to life. And unfortunately they do so often enough to sustain the old dream. But Émile is a willow-the-wisp, who has led many teachers into a conception of their role which could prove disastrous. (p. 3)

Even in the relatively short history of empirical research on teaching, that is, in the second half of the twentieth century, empirical researchers have formulated, advocated, and studied many kinds of teaching. Empirical research began in the 1890s with the pioneering work of Joseph Mayer Rice (1897), but cited in Rice (1913). Since then, researchers have developed and studied a large array of different kinds of process. Rice found no correlation between amount of time used by the teachers for teaching spelling and the achievement of their students on a spelling test. He used this finding to inveigh against "the spelling grind."

Models of the Process of Teaching

A model of the process of teaching is a specific and integrated set of teaching principles and practices for use by teachers who accept the model's implicit or explicit conception of effective teaching. Models often contain special materials and manuals for the teacher using the model. Books describing collections of models of teaching have brought together, defined, and described groups of models. One frequently updated collection – *Models of Teaching* (Joyce, Weil, & Calhoun, 2000) – identified, categorized, and described four "families" of models containing a total of 20 models of teaching shown in Table 4.1.

Researchers have studied some of these varieties of process to see how effective they are in helping students achieve cognitive objectives. Some models are also designed to foster the achievement of *social* and *emotional* objectives.

Two Categories of Models

Probably none of these 20 models of teaching, with one exception discussed below, has been studied and used by more than a few thousand of the 3.5 million U.S. teachers (National Center for Education Statistics, 2000, p. 48).* We make it easier to understand these models by putting them into two categories, namely, Progressive-Discovery-Constructivist models in grades 1–12 and Conventional-Direct-Recitation models.

---

* I was unable to find a statistical survey of how often these or other models are used.

Table 4.1 Four families of teaching models

A. The social family
   Partners in learning
   Role playing
   Jurisprudential inquiry
   Personality and learning styles
   Inquiring on social models

B. The information-processing family
   The basic inductive model
   Concept attainment
   Scientific inquiry and inquiry training
   Memorization
   Synectics (creative thought)
   Learning from presentations
   Developing intellect
   Inquiring on information – processing

C. The personal family
   Nondirective teaching
   Concepts of self
   Inquiring on personal models

D. The behavioral systems family
   Mastery learning and programmed instruction
   Direct instruction
   Learning from simulations

Source: Joyce et al. (2000)

Progressive–Discovery–Constructivist Teaching

The three terms – progressive, discovery, constructivist – represent a chronology. During the first half of the twentieth century in the United States, progressive education's way of teaching was much discussed, but far from widely used. During the third quarter of the twentieth century, discovery teaching was fairly widely considered. In the last quarter of the twentieth century, constructivist teaching received much attention from professors of education. All three kinds of teaching allowed students much freedom to choose their activities according to their interests and prior knowledge of the content, so that for substantial periods of time they could select and carry out their activities on their own initiative, with some guidance, of course, from the teacher.

Constructivism is still, early in the twenty-first century, a thriving concern of educators writing on teaching and teacher education. Among these are Fosnot (1996), Greer, Hudson, and Wiersma (1999), Phillips (1985), and Richardson (1997). Hence, we consider it here. Table 4.2 shows, in an inventory on constructivist practices, a brief example of what constructivist teaching entails.

Hirsch (1996) pointed out that:

constructivism is not only desirable, it is also universal. It characterizes *all* meaningful learning no matter how derived. The nature of one's constructed understanding is normally irrelevant

Table 4.2 The constructivist teaching inventory

A. Community of learners
    1. Interaction to support the challenging and clarifying of ideas (*frequently*; occasionally; seldom) occurs.
    2. Climate of the classroom is (*primarily challenging,* **consistently pushing understanding;** somewhat challenging; primarily non-challenging, does not push understanding).

B. Teaching strategies
    1. Teacher's primary role is to (**facilitate critical student inquiry, not to provide knowledge, skills, and answers**; provide students with knowledge, skills, and answers and to a lesser extent facilitate student critical inventory; provide knowledge, skills, and answers and not to facilitate students' critical inquiry).
    2. Teacher (*intentionally provides*; provides, but not intentionally; does not intentionally provide) students with opportunities for cognitive disequilibrium appropriate for their cognitive understanding.

C. Learning activities
    1. Activities are (seldom; moderately; ***readily***) adaptable to accommodate individual students' interests, needs, and abilities.
    2. Opportunities for both confirming and disconfirming solutions are (**frequently**; occasionally; rarely) provided.

D. Curriculum-assessment
    1. Selection of content for teaching is (almost never; occasionally;*frequently*) based on students' interests, prior knowledge, and/or particular learning needs.
    2. Teacher (***seldom;*** occasionally; frequently) organizes knowledge and skills to be learned in such a way that relationships among them are obvious.

Source: Greer et al., (1999) Sample items from the four categories, with constructivist alternatives in *bold italics*

> to the means by which one constructed it .... Hearing a lecture – in the event that one is understanding it – requires an active construction of meaning. Listening, like reading, is far from being a passive, purely receptive activity. (p. 134)

Despite occasional critical voices, many professional students of teaching favor Progressive–Discovery–Constructivist teaching (PDC). Because constructivist teaching resembles discovery teaching, it probably has the same shortcomings that Ausubel (1963, pp. 139–175) found in discovery teaching. Table 4.3 shows his version of twelve claims of advocates of learning by discovery – claims Ausubel effectively refuted.

After finding none of these ideas to be justified on logical grounds, Ausubel examined the empirical evidence on the effectiveness of learning by discovery. He found that both the short-term and long-term studies provided inadequate evidence because of various methodological shortcomings, such as (a) failure to control for the "Hawthorne Effect" (that is, improvement that occurs merely as a result of the group's perception of their being singled out for special treatment) and (b) absence of a control group.

> An abundantly documented attack on PDC teaching was provided by Kirschner, Sweller, and Clark (2006). They held that PDC teaching, which they labeled "minimally guided instruction," is likely to be ineffective. The past half-century of empirical research on this

Table 4.3 Statements on discovery learning

1. "All real knowledge is self-discovered"
2. "Meaning as an exclusive product of creative, non-verbal discovery"
3. "Subverbal awareness as the key to transfer"
4. "The discovery method in transmitting subject-matter content"
5. "Problem-solving ability as the primary goal of education"
6. "Training in the 'heuristics of discovery'"
7. "Every child a creative and critical thinker"
8. "Expository teaching as authoritarianism"
9. "Discovery organizes learning effectively for later use"
10. "Discovery as a unique generator of motivation and self-confidence"
11. "Discovery as a prime source of intrinsic motivation"
12. "Discovery and the 'conservation of memory'"

Considered unsubstantiated by Ausubel (1963, pp. 139–175)

issue has provided overwhelming and unambiguous evidence that minimal guidance during instruction is significantly less effective and efficient than guidance specifically designed to support the cognitive processing necessary for learning. (p. 76)

Further,

After a half-century of advocacy associated with instruction using minimal guidance, it appears that there is no body of research supporting the technique. In so far as there is any evidence from controlled studies, it almost uniformly supports direct, strong instructional guidance rather than constructivist-based minimal guidance during the instruction of novice-to-intermediate learners. Even for students with considerable prior knowledge, strong guidance while learning is most often found to be equally effective as unguided approaches. Not only is unguided instruction normally less effective; there is also evidence that it may have negative results when students acquire misconceptions or incomplete or disorganized knowledge. (pp. 83–84)

The question of whether the comparisons between PDC and CDR were made on the basis of assessments that would be considered fair to PDC teaching can be considered in terms of the kinds of outcomes measured in the studies cited by Kirschner et al., (2006). Here a typical statement, of a kind that occurs repeatedly, is the following:

Klahr and Nigam (2004), in a very important study, not only tested whether science learners learned more via a discovery versus direct instruction route but also, once learning had occurred, whether the quality of learning differed. Specifically, they tested whether those who had learned through discovery were better able to transfer their learning to new contexts. The findings were unambiguous. Direct instruction, involving considerable guidance, including examples, resulted in vastly more learning than discovery. Those relatively few students who learned via discovery showed no signs of superior quality of learning. (pp. 79–80)

Three replies to the Kirschner-Sweller-Clark argument appeared in the next volume of the same journal. The first (Schmidt, Loyens, van Gog, & Paas, 2007) argued that "Problem-based learning *is* compatible with human cognitive architecture." To this argument Sweller, Kirschner, and Clark responded that problem-based learning "does indeed deemphasize guidance" (p. 115) and thereby increase cognitive load. The second (Hmelo-Silver, Duncan, & Chin, 2007) objected to the characterization of problem-based learning and inquiry learning as unguided discovery learning which increases cognitive load because "Surely the *raison d'etre* of problem-based learning is to deemphasize direct instructional guidance (p. 115). The third (Kuhn, 1962) objected to the Kirschner-Sweller-Clark argument on the grounds that it ignored the problem of *what* to teach by focusing on the less important issue of *how* to teach. The three papers leave the case for conventional-direct-recitation (CDR) teaching still arguable in a way that calls for the sharpening of the issue by means of further experimentation. The case against the desirability of a theory of PDC teaching gets stronger when we consider studies of how most teachers actually teach, as summarized below.

Conventional–Direct–Recitation Teaching

As we've already noted, observations and other evidence suggest that most, by far, of 3.5 million U.S. teachers in grades 1–12 are still using a different model, namely, Conventional-Direct-Recitation (CDR) teaching. CDR designates a contrasting, non-PDC family of teaching. The term "conventional" refers to the ubiquity of CDR teaching since early in the twentieth century in the United States. Investigations of how teachers have taught (for example, Cuban, 1984, 1988; Goodlad, 1984), and are presumably still teaching, have supported the inference that most U.S. teachers practice CDR teaching.

The term "direct" refers to teaching that is teacher-directed and structured, so that the teacher chooses most student activities. The term "recitation" refers to the almost universal pattern whereby the teachers ask questions, and the students respond.

The CDR model flourished in the U.S. throughout the twentieth century, even as alternatives of the PDC kind were widely discussed and advocated. Rosenshine (1987) described CDR teaching as "explicit," and listed its components as follows:

- Begin a lesson with a short statement of goals.
- Begin a lesson with a short review of previous, prerequisite learning.
- Present new material in small steps, with student practice after each step.
- Give clear and detailed instructions and explanations.
- Provide a high level of active practice for all students and obtain responses from all students.
- Guide students during initial practice.
- Provide systematic feedback and corrections.
- Provide explicit instruction and practice for seatwork exercises and, when necessary, monitor students during seatwork.
- Continue practice until students are independent and confident (p. 76).

Rosenshine and Meister (1995, pp. 143–149) identified five variations of "direct teaching":

(a) the teacher-led meaning;
(b) the teacher effectiveness meaning, that is, a set of teaching actions derived from empirical research on teacher effectiveness;
(c) the cognitive strategies meaning, in which researchers developed ways of teaching cognitive strategies – such as summarizing, reading comprehension, and question-generation – and labeled their teaching "direct teaching";
(d) the DISTAR (Direct Instructional Systems in Arithmetic and Reading) meaning, which referred to (i) an explicit step-by-step strategy; (ii) development of mastery at each step in the process; (iii) specific strategy corrections for student errors; (iv) gradual transition from teacher-directed activities toward independent work; (v) use of adequate and systematic practice through a range of examples of the task; (vi) many classroom settings in which instruction is led by the teacher, particularly settings in which the teacher lectures and the students sit passively, and (vii) the undesirable-teaching meaning, which refers to direct teaching as "authoritarian," "regimented," "fact accumulation at the expense of thinking-skill development," and "focusing on tests." (Rosenshine & Meister, 1995, pp. 143–149);
(e) Another usage of "direct instruction," referred to a model developed by Engelmann (1980) that emphasizes the use of carefully prepared lessons, designed around a highly specified knowledge base and a well-defined set of skills for each subject. A central element of the theory underlying Direct Instruction is that clear instruction eliminates misinterpretations and can greatly improve and accelerate learning. (American Institutes for Research, 1999, p. 63);
(f) Cuban (1988) used the term "teacher-centered instruction" as his label for what was essentially CDR:
A cumbersome phrase, teacher-centered instruction tries to capture a common form of instruction where teachers generally teach to the whole group of students in a class, show high concern for whether students are listening, concentrate mostly on subject matter and academic skills, and, in general, control what is taught, when, and under what conditions. (p. 27);
(g) Joyce et al. (2000) provided another description:
The most prominent features [of CDR teaching] are an academic focus, a high degree of teacher direction and control, high expectations of pupil progress, a system for managing time, and an atmosphere of relatively neutral affect. (p. 338);
(h) Burns (1984) summarized "descriptors of direct instruction gleaned from the recent reviews of research on teaching" (p. 106). His compilation is shown in Table 4.4.

In short, CDR teaching is relatively highly structured, and the student plays a seemingly, but not actually, passive role along lines established by the teacher. Of the 20 models described by Joyce et al. (2000), CDR is probably by far the most widely used. Cuban's history, *How Teachers Taught* (1984) described CDR as the almost universal process of teaching in the United States between 1890 and 1980. Goodlad's *A Place Called School* (1984), in describing the process of teaching

Table 4.4 Snapshot data: rank order of activities by probability of students having been observed participating in each at any particular moment

Early elementary activity	%	Upper elementary activity	%
Written work	28.3	Written work	30.4
Listening to explanations/lectures	18.2	Listening to explanations/lectures	20.1
Preparation for assignments	12.7	Preparation for assignments	11.5
Practice/performance – physical	7.3	Practice/performance – physical	7.7
Use of AV equipment	6.8	Use of AV equipment	5.5
Reading	6.0	Reading	5.3
Student non-task behavior – no assignment	5.7	Student non-task behavior – no assignment	4.9
Discussion	5.3	Discussion	4.8
Practice/performance – verbal	5.2	Practice/performance – verbal	4.4
Taking tests	2.2	Taking tests	3.3
Watching demonstrations	1.5	Watching demonstrations	1.0
Being disciplined	0.5	Being disciplined	0.4
Stimulation/role play	0.2	Stimulation/role play	0.3
Junior high activity	**%**	**Senior high activity**	**%**
Written work	21.9	Written work	25.3
Listening to explanations/lectures	20.7	Listening to explanations/lectures	17.5
Preparation for assignments	15.9	Preparation for assignments	15.1
Practice/performance – physical	14.7	Practice/performance – physical	12.8
Use of AV equipment	5.5	Use of AV equipment	6.9
Reading	4.2	Reading	5.8
Student non-task behavior – no assignment	4.2	Student non-task behavior – no assignment	5.1
Discussion	4.1	Discussion	4.5
Practice/performance – verbal	3.6	Practice/performance – verbal	2.8
Taking tests	2.8	Taking tests	1.9
Watching demonstrations	1.5	Watching demonstrations	1.6
Being disciplined	0.2	Being disciplined	0.1
Stimulation/role play	0.2	Stimulation/role play	0.1

Source: Burns (1984, p. 107)

observed in use by 1,017 teachers (far more than any other observational study of teaching), made it very likely that U.S. teachers are still teaching according to the CDR model. Thus, we can safely conclude that the CDR model survived throughout the twentieth century, while PDC-style alternatives were much advocated, but not used nearly as often.

Empirical Studies of the Process of Teaching

Empirical studies of teaching processes have used a variety of methods: historical, stenographic, and observational methods.

A Historical Study

Cuban's *How Teachers Taught* (1984) provided historical information on what kind of teaching occurred in U.S. classrooms during the twentieth century. Cuban used an array of non-observational kinds of evidence:

> how classroom furniture and space were arranged, what manner of grouping for instruction the teacher used (whole class, small groups, and so forth), classroom talk by teacher and students, activities that students and teacher engaged in (recitation, discussion, reports, tests, film, lecture, and so forth), and the amount of physical movement allowed the student within the classroom. These categories were visible signs of how teacher-centered the class was. As it turned out, the degrees of difference over the decades were quite small (pp. 28–29).

From these kinds of evidence, Cuban inferred the following:

> To the question—how did teachers teach?—answers can now be drawn from a substantial body of evidence, direct and contextual, from 1900 clearly showing what the central teaching tendency was and what variations of that dominant strain existed. Precision in methodology and sampling of historical sources were limited. However, the collection of almost 7,000 different classroom accounts, and results from studies in numerous settings, revealed *the persistent occurrence of teacher-centered practices since the turn of the century* [italics added]—at the sizable risk of dulling a reader's sensibilities by presenting similar patterns and numbers. This historical inquiry into classroom instruction and the imprecise responses were in the spirit of one researcher [Tukey, 1962], who said, "Far better an approximate answer to the right question, which is often vague, than an exact answer to the question which can always be made precise." (p. 238)

Cuban inferred that, from 1890 to 1980, despite widespread and intense discussion, advocacy, and rationalizing rhetoric favoring PDC teaching – the majority of U.S. teachers continued to use CDR. In summary, Cuban (1984) wrote:

> Drawn from a large number of varied sources in diverse settings, over nearly a century, the data show striking convergence in outlining a stable core of teacher-centered instructional activities in the elementary school and, in high school classrooms, a remarkably pure and durable version of the same set of activities. (p. 238)

Observational Studies

Researchers' observations of classrooms were rare until about 1960. Some of the better known observational studies were done by Bellack, Kliebard, Hyman, & Smith (1966), Hoetker and Ahlbrand (1969), Mehan (1979), and Goodlad (1984). Although the researchers used different terminologies, the overall portrait suggested great uniformity in how teaching went on throughout the twentieth century for virtually all grade levels and subject matters. CDR teaching apparently prevailed.

Bellack et al. (1966)

Within the uniformity of the CDR model of teaching, we find what Bellack et al. (1966) called, in the title of their extraordinarily thorough monograph, *The Language of the Classroom.* This "language" was derived from a study of the way in which 15 teachers taught a unit on international trade in 15 high school classes. The researchers analyzed their teaching in terms of a series of cycles, or sets of exchanges, between the teacher and the students.

In their complete form, these cycles consist of various combinations of what *Language's* authors called "moves," namely, "structuring," "soliciting," "responding," and "reacting," defined in Table 4.5. The cycle is then repeated in a variety of ways, with one or more of its components either repeated or omitted within a cycle.

The findings of Bellack et al., were supported by the non-observational work of Smith, Meux, Coombs, Eierdam, & Szoke. (1962) in "A Study of the Logic of Teaching":

> In its normal course, a discussion in progress exhibits a characteristic development. Certain forms of utterances are used to enjoin or invite immediate reply; other forms are conventionally understood to forestall or prohibit immediate response [compare Bellack et al.'s *Structuring*]. A direct question, addressed either to a given person or to the group at large, conventionally demands some kind of responding action on the part of the individual or group addressed [compare Bellack et al.'s *Soliciting*]. A rhetorical question, on the other hand, is commonly understood to be uttered for its dramatic or rhetorical effect, but some do serve to trigger discussion. When a reply is made to a direct question [compare Bellack et al.'s *Responding*], it is also a convention that the reply itself be acknowledged in some way, at least by word or gesture if not by further responding commentary or questioning [compare Bellack et al.'s *Reacting*]. (p. 12)

Some of these moves might be omitted in a teaching cycle, as shown in Table 4.6.

A direct question, addressed either to a given person or to the group at large, conventionally demands some kind of responding action on the part of the individual or group addressed [com-

Table 4.5 Definitions of structuring, soliciting, responding, and reacting

1.1. *Structuring* (STR). Structuring moves serve the function of setting the context for subsequent behavior by (1) launching or halting-excluding interactions between teacher and pupils, and (2) indicating the nature of the interaction in terms of the dimensions of time, agent, activity, topic and cognitive process, regulations, reasons, and instructional aids. A structuring move may set the context for the entire classroom game or a part of the game.

1.2. *Soliciting* (SOR). Moves in this category are intended to elicit (a) an active verbal response on the part of the persons addressed; (b) a cognitive response, e.g., encouraging persons addressed to attend to something; or (c) a physical response.

1.3. *Responding* (RES). Responding moves bear a reciprocal relationship to soliciting moves and occur only in relation to them. Their pedagogical function is to fulfill the expectation of the soliciting moves and is, therefore, reflexive in nature. Since solicitations and responses are defined in relationship to each other, there can be no solicitation that has not intended to elicit a response, and no response that has not been directly elicited by a solicitation.

1.4. *Reacting* (REA). These moves are *occasioned* by a structuring, soliciting, responding, or a prior reacting move, but are not directly elicited by them. Pedagogically, these moves serve to modify (by clarifying, synthesizing, or expanding) and/or to rate (positively or negatively) what is said in the move(s) that occasioned them. Reacting moves differ from responding moves: while a responding move is always directly elicited by a solicitation, *preceding moves serve only as the occasion for reactions.* For example, the rating by a teacher of a student's response is designated a reacting move; that is, the student's response is the occasion for the teacher's rating reaction but does not actively elicit it.

Source: Bellack et al. (1966, pp. 16–19)

Table 4.6

1.	STR				
2.	STR	SOL			
3.	STR	REA			
4.	STR	REA	REA ...		
5.	STR	SOL	RES ...		
6.	STR	SOL	RES	RES ...	
7.	STR	SOL	REA		
8.	STR	SOL	REA	REA ...	
9.	STR	SOL	RES	REA	
10.	STR	SOL	RES	REA	REA ...
11.	STR	SOL	RES	REA	RES ...
12.	STR	SOL	RES	REA	RES ... REA ...
13.	SOL				
14.	SOL	RES			
15.	SOL	RES			
16.	SOL	REA			
17.	SOL	REA	REA ...		
18.	SOL	RES			
19.	REA	REA	REA		
20.	REA	REA ...			
21.	SOL	RES	REA	RES ...REA ...REA ...	

STR structuring, *SOL* soliciting, *RES* responding, *REA* reacting

pare Bellack et al.'s *Soliciting*]. A rhetorical question, on the other hand, is commonly understood to be uttered for its dramatic or rhetorical effect, but some do serve to trigger discussion. When a reply is made to a direct question [compare Bellack et al.'s *Responding*], it is also a convention that the reply itself be acknowledged in some way, at least by word or gesture if not by further responding commentary or questioning [compare Bellack et al.'s *Reacting*]. (p. 14)

The wide applicability of the kind of analysis of teaching developed by Bellack and those who reported subsequently was described by Gage (1979) as follows, on the basis of a report by Bellack (1976):

> During the following decade, that analysis was found useful in about thirty-five related studies describing teaching at every grade level from elementary through college; in subjects as varied as reading, arithmetic, mathematics, science, teaching, and nursing; in six other countries (Sweden, Finland, Australia, Germany, Canada, and Japan); and in such varied settings as individualized instruction, mathematics in "open" elementary school classrooms, and early education programs. In all these studies, the pattern of structuring, soliciting, responding, and reacting was found to occur in clearly defined ways. (pp. 273–274)

Hoetker and Ahlbrand (1969)

In a study of nine junior high-school English teachers, Hoetker and Ahlbrand (1969) found that those classes "behaved very precisely [as shown in Table 4.7] according to Bellack et al.'s 'rules.'" According to Nuthall and Snook (1973),

Table 4.7 Comparisons between selected mean measures of classroom verbal behavior in Bellack, et al. (1966) and Hoetker and Ahlbrand (1969)

Measure	Bellack et al. (1966)*	Hoetker and Ahlbrand (1969)
A. Percentage of teacher talk, moves	61.7	65.7
B. Percentage of teacher talk, lines of typescript	72.1	74.5
C. Distribution of teacher moves, as percentage of all moves		
STRUCTURING	4.8	3.6
SOLICITING	28.8	32.3
RESPONDING	3.5	1.8
REACTING	24.3	27.0
D. Distribution of pupil moves, as percentage of all moves		
STRUCTURING	0.4	0.3
SOLICITING	4.4	2.0
RESPONDING	25.0	30.4
REACTING	5.7	1.1
E. Distribution of teacher moves, as percentage of total lines of typescript		
STRUCTURING	14.5	22.4
SOLICITING	20.3	20.6
RESPONDING	5.0	4.3
REACTING	24.8	31.4
F. Distribution of pupil moves, as percentage of total lines of typescript		
STRUCTURING	3.0	3.4
SOLICITING	2.5	1.2
RESPONDING	15.6	13.1
REACTING	5.1	0.6
G. Percentage of teacher questions calling for memory processes	80.8*	87.9

Source: Hoetker and Ahlbrand (1969, p.146)
*Estimated from data in Bellack et al. (1966, pp. 74–75)

> The evidence gathered by Hoetker and Ahlbrand (1969) strongly suggests that this class-room language game has had a long and persistent history. Records of observational studies from the turn of the nineteenth century indicate that the game has not changed substantially in approximately 60 years. (p. 52)

Mehan (1979)

Studying a single class of pupils in grades 1–3 for a whole school year, Mehan (1979) reported that the class exhibited three of Bellack et al.'s four components of a teaching cycle: (a) "initiation" (compare the structuring and soliciting of Bellack et al.), (b) "responding," typically by students (compare the responding of Bellack et al.), and (c) "evaluating" (compare the reacting of Bellack et al.).

Goodlad (1984)

It is, of course, desirable to have large-scale studies on which to base a formulation of what occurs in CDR teaching. Such a study was provided by Goodlad's tour de force, *A Place Called School* (1984), a report on how 1,017 teachers taught. Goodlad's staff observed the 1,017 teachers in a representative sample of 38 schools (13 senior high schools, 12 junior high or middle schools, and 13 elementary schools) in seven states in all regions of the U.S. The observers were carefully trained and retrained until their results showed a high degree of agreement between observers. The observers used a modification of the observation system developed by Stallings and Kaskowitz (1974).

As Goodlad (1984, p. 18) wrote, "The schools we studied differ in location, size, characteristics of the student population, family incomes, and other ways. Nonetheless … they shared many similarities, particularly in modes of teaching and learning" (p. 18). We examine *A Place Called School* to see how its descriptions of teaching support or differ from those of the much smaller studies. The book's fourth chapter, "Inside Classrooms," reports on the process of teaching observed in the classrooms studied.

Table 4.3 shows that the high-ranking "Listening to Explanations/Lectures" represents the counterpart in student activity of the teacher's "structuring" and "soliciting" components of the formulation by Bellack et al. (1966). The nearly-as-high rank of the pupil's "practice/performance-verbal" represents the counterpart of the "responding" component of Bellack et al. (1966). Sirotnik (1983), reporting on the data from Goodlad's study, wrote:

> Providing corrective feedback in combination with additional information designed to help students understand and correct their mistakes *is almost nonexistent* [italics added]. In fact, reinforcement of any kind *is rarely noticed* [italics added], whether in the form of specific task-related acknowledgement and praise or general support and encouragement. (p. 20)

These findings represent a departure from the findings of Bellack et al., concerning teachers' reacting, in that Goodlad's teachers' reactions betoken mere general, non-specific acknowledgment of students' responses.

In summary, the typical classroom patterns – as reported in *A Place Called School* – consisted of the teacher's (a) explaining or lecturing to the whole class or to a single student (compare structuring), (b) asking direct, factual questions on the subject matter, or monitoring students (compare soliciting), (c) the students' ostensibly listening to the teacher and responding to teacher-initiated interaction (compare responding) and (d) the teacher's providing non-specific acknowledgement of students' responses.

We can sort Sirotnik's descriptions according to the categories of Bellack et al. (1966). Such sorting indicates that the findings of *A Study of Schools* (Sirotnik, 1983) agree well with the structuring-soliciting-responding-reacting analysis of classroom teaching constructed by Bellack et al. (1966), and confirmed by Hoetker and Ahlbrand (1969) and Mehan (1979).

Working with the data from Goodlad's study of 1,017 teachers, Sirotnik (1983) summarized some of the findings, in ways that can be aligned with the categories of Bellack et al. (1966):

(a) *Structuring.* "Nearly 70 percent of the total class time involves verbal interaction, or 'talk' .... Less than a fifth of the time involves student talk." (Sirotnik, 1983, p. 20)

(b) *Soliciting.* "[W]e find that barely five percent of the instructional time is spent on direct questioning — questioning which anticipates a specific response like 'yes,' 'no,' 'Columbus,' or '1492.' Less than 1 percent of that time is devoted to open questions which call for more complex cognitive or affective responses." (Sirotnik, 1983, p. 20)

(c) *Responding.* "[T]he most frequently occurring single interaction, is one of students responding to the teacher .... This occurs roughly 15 percent and 10 percent of the time at the elementary and secondary levels, respectively." (Sirotnik, 1983, p. 20)

(d) *Reacting.* "[L]ess than 5 percent of teacher's time is spent responding to students, which, as will be seen shortly, is less than the percentage of time students are observed initiating interaction with the teacher .... Providing corrective feedback in combination with additional information designed to help students understand and correct their mistakes is almost nonexistent. In fact, reinforcement of any kind is rarely noticed, whether in the form of specific task-related acknowledgement and praise or general support and encouragement." (Sirotnik, 1983, p. 20)

In summary, according to Sirotnik (1983, pp. 20–21),

The model classroom pattern consists of the teacher's (a) [structuring, that is,] explaining or lecturing to the whole class or to a single student; (b) [soliciting, that is,] then asking direct, factual questions on the subject matter, or monitoring students; (c) [responding, that is,] the students ostensibly listening to the teacher or responding to a teacher-initiated interaction; and (d) [reacting, that is,] mere acknowledgement of students' responses rather than the teacher's indicating whether the student's response was correct and going on to other relevant comments.

So we can sort these quotations from Goodlad's large-scale study according to the Bellack et al. categories. Doing so indicates that the findings of *A Place Called School* (Goodlad, 1984; Sirotnik, 1983) – concur substantially with the structuring-soliciting-responding-reacting analysis of classroom teaching constructed by Smith et al. (1962), and Bellack et al. (1966), Hoetker and Ahlbrand (1969), and Mehan (1979).

Similarity of the Bellack Model to Computer-Assisted Instruction Frames

Apparently, the CDR conception of the teaching process is not unique to classroom teaching, but has more general applicability. It is similar to the conception of instruction discernible in the "frames" of computer-assisted instruction. In the following comparisons, material quoted from *Computer Based Instruction* (Alessi & Trollip, 1985) is italicized:

(a) Such frames begin with "*a piece of instruction,*" ranging in size from a sentence to a long paragraph, concerning some aspect of the subject matter being taught [compare Bellack et al.'s "structuring"];

(b) The next part of the frame is a "*question or problem addressed to the student*" [compare Bellack et al.'s "soliciting"];

(c) The student then *"replies"* to this question or problem [compare Bellack et al.'s "responding"];

(d) The cycle is completed by giving the student an *"evaluation of response"* ("Right" or "Wrong") [compare Bellack et al.'s "reacting"].

The computer-assisted instruction frames were developed by researchers formulating instruction via a medium, a computer, radically different from the classroom teacher. Its similarity to the Bellack et al., formulation supports the proposition *that the Bellack model embodies something profoundly fundamental in the nature of teaching.*

The Generalizability of the CDR Model

Classroom observations have suggested that teaching in U.S. schools consists typically of a series of cycles, or exchanges between the teacher and students, of the kind originally described by Smith et al., (1962) on logical grounds and subsequently by Bellack et al., (1966), Hoetker and Ahlbrand (1969), Mehan (1979), and, by inference, Sirotnik (1983), and the frames of computer-assisted instruction.

The cycle repeats – sometimes with modifications such as those shown in Table 4.5, which do not change the essential character of the process: *Structuring* leads to *soliciting* which leads to *responding* which leads to *reacting.*

CDR Teaching Across Nations

So far, we have presented evidence on CDR teaching only from U.S. studies. Does it also predominate in other countries, in other cultures? Smith et al., (1962) assumed it does:

> Teaching is here assumed to be a social phenomenon, *fundamentally the same from one culture to another* [italics added]. It has its own elements, forms, regularities, and problems. It takes place under what seems to be a relatively constant set of conditions—time limits, authority figures, student ability limits, institutional structures, etc. (p. 2)

The IEA Classroom Environment Study (Anderson, Ryan, & Shapiro, 1989) – conducted under the auspices of the International Association for the Evaluation of Educational Achievement (IEA) – studied precisely this question. The ten participating political entities were Australia, Canada (Ontario, English), Canada (Ontario, French), Canada (Quebec), Hungary, Israel, South Korea, the Netherlands, Nigeria, and Thailand.

Classroom observation while teaching is underway provides the best evidence on the process of teaching – but it is also expensive. So only five of the regions provided such evidence: Australia, Canada (Ontario, English), Canada (Ontario, French), Canada (Quebec), and Hungary. The subject taught and therefore observed in four of these five political entities was mathematics (Ryan, Hildyard, & Bourke, 1989, p. 48). In Hungary, physics was taught. All five of these political entities used the same observation schedules, trained observers the same way, and obtained substantial agreement between observers. Mandeville (1989) summarized the teaching in these five political entities as follows:

> First, most lessons were divided into two or more segments, lasting an average of 10 to 15 minutes each. Second, during review, oral practice, lecture, and discourse segments, teachers typically assumed a very directive, interactive role with their students. During written or laboratory seatwork segments, however, role differentiation between teachers in many countries was evident. *In many seatwork segments, teachers engaged in what may be termed "teaching" (that is, they explained concepts and skills to their students, asked questions and reacted to their responses).*
>
> *Third, teachers generally spent a great deal of time providing explanations, asking questions, and reacting to answers to those questions during the entire lesson. These behaviors were fairly common in review, oral practice, lecture, and discourse segments* [italics added]. (p. 145)

The findings italicized are extraordinarily similar to those reported by studies in the U. S. (Bellack et al., 1966; Cuban, 1984; Goodlad, 1984; Hoetker & Ahlbrand, 1969; Mehan, 1979). They suggest that CDR teaching of this kind occurs not only in U.S. classrooms but in other parts of the world as well.

Stigler and Hiebert (1999) reported another study of international similarities and differences in the process of teaching. It was titled. The Third International Mathematics and Science Study (TIMSS). They studied videotapes of eighth-grade mathematics as it was taught in the United States (81 classrooms), Germany (100 classrooms), and Japan (50 classrooms). But their rich treatment of teaching and teacher education did not focus on cycles of teaching – that is, the recurring sequences of teacher and student moves exemplified by the structuring-soliciting-responding-reacting identified in the studies cited above. So their results cannot be compared with those of the foregoing studies.

CDR Teaching Across Subject Matters

The studies of CDR teaching in the U. S. dealt with subject matter that is expressed primarily in words – namely, international trade (Bellack et al., 1966), English (Hoetker & Ahlbrand, 1969), and the varied content of a class for combined grades 1–3 (Mehan, 1979), So we might question whether the CDR model also prevails for largely nonverbal subjects, such as mathematics and science. Here again, *The IEA Classroom Environment Study* is helpful, because the subject taught in four of the five participating political entities – Australia, Canada (Ontario, English), Canada (Ontario, French), Canada (Quebec) – was mathematics. In the fifth (Hungary), it was physics.

Since the IEA study did not report that their findings differed from those for verbal subjects, as taught outside the United States, we can reasonably infer that the same teaching processes were observed for mathematics and physics.

So the IEA Study's findings further support the ubiquity of CDR teaching. The IEA study dealt with many aspects of teaching other than its process, such as content taught, student characteristics, school variables, and variables in students' homes.

In summarizing what the IEA Study's observers learned about the structure of lessons, Mandeville (1989) reported:

> [T]eachers generally spent a great deal of time providing *explanations, asking questions, and reacting* [italics added] to answers to those questions during the entire lesson. These behaviors were fairly common in review, oral practice, lecture, and discourse segments. Although not as predominant, these behaviors also were evident while students worked on assigned tasks at their seats. (p. 145)

The close resemblance of this description in the five U.S. observational studies (Bellack et al., 1966; Cuban, 1984; Goodlad, 1984; Hoetker & Ahlbrand, 1969; Mehan, 1979) suggests that teachers of both verbal and highly non-verbal subject matter follow the same conventional model of teaching.

The Reader's Memory

Perhaps the most persuasive basis for accepting the prevalence of CDR teaching in U.S. schools is not to be found in research of any kind, however valid. Rather, its well-established ascendancy may reside in the memories of this book's readers, recalling their own classes on their way to earning a high school diploma. Those recollections assure us that we have not erred in selecting CDR teaching as the model for our theory and our attempt to explain its effectiveness. A comment from Goodlad (1984) should ring a bell for many readers:

> The classrooms we observed were more like than unlike those in the old images so many of us share. Usually we saw desks or tables arranged in rows, oriented toward the teacher at the front of the room. Instructional amenities such as library corners, occasionally present in elementary classrooms, were rarely observed in secondary classes. The homelike chairs and rugs sometimes seen in primary classes rapidly became rare with upward progression through the grades.... The central focus is on teachers' pedagogical practices – grouping, individualizing, using time, making decisions. (p. 94)

Present Status of the Search for the Prevalent Model of Teaching

When all is said and done, we must question our conclusions about CDR. We have used inductive logic, in compiling evidence on its prevalence. That is, if repeated observations yield the same result, namely, the prevalence of CDR, can it be regarded as *proven* to be true?

Reasons for Suspending Judgment

We must argue against ourselves because of long-established questions about inductive logic and the extreme amount of evidence not yet examined.

Inductive Logic's Inadequacy

As philosophers (Hume, 1758; Popper, 1965) famously pointed out, inductive logic cannot prove the inevitability of a conclusion. No matter how much evidence – stenographic, historical, observational, or remembered personal experience – is accumulated, we cannot *logically* conclude, on the basis of inductive logic, that we have identified the model of teaching used by most teachers in the United States and elsewhere. In other words, just because the sun has risen every morning in the East for untold millennia, we cannot *logically* prove that it will always do so. Just because we have accumulated much evidence that CDR teaching prevails, we have not proved that it does prevail.[†]

Inadequacy of the Evidence

Also, the common sense of anyone familiar with the abundance of the literature of research on teaching should dissuade us from any firm conclusion. For example, the foregoing body of evidence is not based on thorough analyses of how teaching varies as a function of grade level, subject matter, students' SES, students' scholastic aptitude, teachers' years of experience, and other factors, including combinations and permutations of these factors.

Similarly, the existence of professional organizations and their journals – each devoted to advancing research on teaching and teacher education on particular grade levels, such as elementary (*Elementary School Journal*) or secondary (*School Review*) – should make us cautious before generalizing about the prevalent model of teaching. Those journals carry many papers advancing proposals for the improvement of teaching in their grade levels.

The same caution should follow from the work by specialists in the teaching of particular school subjects (see Brophy, 2001), such as English (for example, *The English Journal*), history (for example, *Teaching Social Studies in an Age of Crisis*), mathematics (for example, *The Arithmetic Teacher*), or science (for example, *Journal of Research in Science Teaching*).

[†]Lakatos (1968) edited a volume of papers by philosophers that raised many questions about Popper's formulations of inductive logic. The present acceptance of Popper's position may need revision in the light of those papers.

Present Conclusion as a Conjecture

Given such caveats, and following the advice of philosopher of science Karl Popper, we regard our present decision to focus our theory on CDR teaching as a *conjecture* (Popper, 1965). Pending the development of bases in deductive logic, supported by inductive evidence, for formulating the model of teaching that will serve as the focus of a theory of teaching, our decision is a matter of convenience – a jumping to a conclusion. It can be accepted as long as it survives rigorous attempts at its *refutation* (Popper, 1965). Such attempts could take the form of strong evidence, supported by deductive logic, that some other model of teaching is more prevalent than CDR teaching.

The CDR model of teaching comprises those models in which the teaching is highly structured, and the student plays a role that only seems to be passive. The apparent passivity is likely to conceal the students' vigorous cognitive processes along lines set by the teacher.

Of the models described by Joyce et al., (2000), we shall proceed on the questionable assumption that none except CDR has in all likelihood achieved predominant use by the millions of grade 1–12 teachers in U.S. elementary and secondary schools. The CDR model has persisted throughout the twentieth century, while alternatives of the PDC kind were widely championed but not used nearly as often. CDR teaching, presumably, will prevail in at least the early decades of the twenty-first century.

Why the Persistence of CDR Teaching?

John Dewey and other influential thinkers led the revolutionary movement that envisioned progressive education. Classroom observers did not expect to find CDR teaching's prevlence because of work by the authors and advocates of the many other models of teaching that were described by Joyce et al., (2000) and older volumes such as *The Passing of the Recitation* (Thayer, 1928). The "passing" that Thayer and many others predicted and hoped for has been belied by research on the teaching that was actually going on–including historical descriptions not based on observations (Cuban, 1984; Lamm, 1976, pp. 1–6) and observation-based descriptions (Bellack et al., 1966; Goodlad, 1984; Hoetker & Ahlbrand, 1969; and Mehan, 1979).

Despite Dewey's efforts to overthrow CDR teaching, a non-Deweyan way of teaching persisted from teacher to teacher, from grade level to grade level, from subject matter to subject matter, from region to region, from decade to decade, and from nation to nation. Yet what we have seen and what we continue to see is little change in the way teaching goes on. Cuban (1982) expressed a common dismay:

> What nags at me is the puzzling durability of this teaching at all levels of schooling but most clearly and uniformly at the high school, decade after decade, in spite of changes in teacher preparation, students' knowledge and skills, and continuous reform fervor to alter this form of instruction. (p. 28)

Sirotnik (1983), on the basis of observations in 1,017 classrooms, reported the same uniformity:

> What we have seen and what we continue to see in the American classroom—the process of teaching and learning—appears to be one of the most consistent and persistent phenomena known in the social and behavioral sciences. To put it succinctly, the "modus operandi" of the typical classroom is still didactics, practice, and little else (pp. 16–17).

Many explanations of this persistence suggest themselves:

"Inheritance" of CDR. The twentieth century's generations of teachers were taught in the CDR way when they themselves were children and adolescents. As Goodlad (1984) reported, after his observational survey of teachers, "teachers teach very much as they were taught through sixteen or more years of classroom life" (p. 306). When some of these students later entered teacher education programs, the new influences were evidently not strong enough to break the hold of their personal experience as students.

This general view was shared by Wallen and Travers (1963). They concluded that teachers were influenced in their teaching "pattern" more by their own teachers than by the teacher education program they had undergone. As Eisner (2004) put a related point:

> Schools have a special difficulty in changing their [own] nature. Part of this difficulty stems from the fact that all of us have served an apprenticeship in them — and from an early age. Indeed, teaching is the only profession I know in which professionalization begins at age 5 or 6. Students, even those of so tender an age, learn early what it takes to "do school." *They learn early what a teacher does in a classroom* [italics added]. They learn early how they must behave in order to get on. (p. 648)

In his attempt to explain the persistence, Nuthall (2005, p. 895) invoked the concept of societal culture by quoting Stiegler and Hiebert (1999, p. 87): "Despite changing teacher education programs and many attempts to reform teaching methods, the core of the ritual remains largely unchanged, sustained by a 'stable web of beliefs and assumptions that are part of the [wider] culture.'"

CDRs Apparent Adequacy. Until recently, CDR teaching seemed to work well enough for the middle-class students who made up the majority of U.S. students. Cuban (1984) saw that reality as the cause of the "persistence of the inevitable" namely, CDR teaching. If U.S. teaching were as grossly inadequate as some critics hold, most U.S. adults would be incompetent as citizens and as workers, and most U.S. students would be markedly inferior to those in other industrialized countries in their achievement test scores.

Yet the U. S. polity, in which the citizens taught *à la* CDR have participated, is one of the most democratic, enlightened, and demanding of cognitive skills among the nations of the world. The U.S. economy in which they work is among the world's most efficient and productive. Berliner and Biddle (1995) brought together abundant evidence against the "attack" on U.S. public schools.

Excessive Demands of Alternatives to CDR. Progressive education, open education, constructivist education, and other kinds of proposed improvements, simply demand more than what most teachers can provide by way of skill, stamina, and conviction. Schools may resist teachers' departures from the uniformity that has been so widely observed for so long. Cremin (1964), a Pulitzer prize-winning historian of U.S. education, illuminated some of the reasons why PDC was never widely practiced:

> From the beginning, progressivism cast the teacher in an almost impossible role: he was to be an artist of consummate skill, properly knowledgeable in his field, meticulously trained in the art of pedagogy, and thoroughly imbued with a zeal for social improvement. It need hardly be said that here as elsewhere on the educational scene of the [eighteen] nineties, the gap between real and ideal was appalling. (p. 168)

"Open education" – a kind of progressive education that was moderately popular for a few decades, especially in England – also asked too much of teachers. Rothenburg (1989) cited evidence that its teachers were not prepared for it, did not integrate its elements into a unified approach, and omitted important parts of the curriculum.

Giaconia and Hedges (1982) carried out a meta-analysis, a quantitative synthesis, of the results of 153 studies that compared open education with CDR in effectiveness. They concluded that although open education did foster more favorable student attitudes towards school, it did not help students achieve cognitive objectives more effectively than CDR.

Weak Effect of Computers in Classrooms. Computers improved markedly and appeared more frequently in schools during the last quarter of the twentieth century. Many educators expected radical changes in the classrooms as a result. Nevertheless, CDR teaching maintains its predominance into the twenty-first century.

Cuban (2001) eloquently titled his book on computers in the classroom, *Oversold and Underused.* In "Summing Up," he wrote:

> The introduction of computers into classrooms in Silicon Valley [the California region in which the computer industry is especially strong] had a number of unexpected consequences. They are:
>
> — Abundant availability of a "hard" infrastructure (wiring, machines, software) and a growing "soft" infrastructure (technical support, professional development) in schools in the late 1990s has not led, as expected, to frequent or extensive teacher use of technologies for tradition-altering classroom instruction.
> — Students and teachers use computers and other technologies more at home than at school.
> — When a small percentage of computer-using teachers do become serious or occasional users, they – contrary to expectations – largely maintain existing classroom practices rather than alter customary practices. (pp. 170–171)

Accordingly, when teacher education programs sought to train teachers so they could capitalize on computers, their influence proved to be too weak to do so. Cuban looked at the past and present place of computers in teaching. He found what seemed like a discouragingly small amount of progress since the manifesto by Suppes (1966) that painted a much brighter future.

But attempts to forecast where information technology will take teaching still appear. One crescendo of advocacy was presented by Zuckerman (2005) in a one-page editorial that envisioned an awesome expansion of teaching possibilities:

It means a teacher can take the class around the world electronically to look at the development of civilizations in Egypt, Greece, Rome, Latin America. A Spanish class in Idaho can talk to students in Bilbao. It means linking biology students in Chicago with a researcher at a microscope in San Francisco, history students with a curator at the National Portrait Gallery, technology students with the National Air and Space Museum in Washington. (p. 68)

But D. K. Cohen (1988) added four additional possible reasons for the persistence of CDR:

The Great Autonomy of U.S. School Systems. The United States has no national curriculum and no prescribed teaching practices. This situation makes it easy for teachers to resist change.

The Conditions of Teaching. Teachers have to use a curriculum they did not develop, according to an imposed schedule, under heavy workloads including extracurricular activities, with low pay and little prestige. These conditions weaken teachers' motivation to adopt new ways of teaching. But, as Cohen pointed out, this reasoning does not explain why teachers in private schools and colleges taught in much the same way as teachers in public schools – even though their teaching loads were lighter, pay was better, and they had more freedom to develop curriculum.

Flaws in Reform. Reform efforts had to cope with inadequate resources and school administrators' insensitivity to teachers' needs. Yet even when such conditions improved greatly, Cohen noted, efforts to change teaching had little success.

Weak Incentives for Change. Because public schools had little competition, their administrators and teachers were not motivated to change.

In Cohen's (1988, p. 36) view, his own explanations "do not seem sufficient to explain the glacial pace of change in teaching". Even when these obstacles to change were absent, as in private schools and colleges, teaching remained for the most part unchanged.

Is Progressive Education Still Around?

Despite all the evidence to the contrary, some writers continue to assert that progressive education – a major component of the PDC model – is not only widespread, but has caused American schools to "go wrong" (Evers, 1998) and become altogether unlike "the schools we need" (Hirsch, 1996). These writers seem to contradict what we have characterized as the predominant style of teaching in U.S. schools and thus our conclusion that progressive education is absent from U.S. classrooms.

Evers (1998, pp. 1–2) wrote that "School reformers today are still trying to put into effect the turn-of-the-century progressive education ideas of John Dewey and others – often these days under the banner of 'discovery learning.'" Progressive education, he said, vanished for a few decades beginning in the mid-1950s "But progressive education came back and is quite influential today in its contemporary incarnation of discovery learning."

Similarly, Hirsch (1996) wrote:

> The *anti-subject-matter viewpoint* [italics added] of *Cardinal Principles* [a 1918 publication of a Commission on the Reorganization of Secondary Education] has dominated the training and certification of teachers in our teacher-training schools since the 1930s, that is, during the entire working lives of all persons now teaching in our schools. (p. 49)

We should note that Evers referred to what "school reformers," not teachers, are doing. And Hirsch similarly referred to "the training and certification of teachers in our teacher training schools," but not to how and what teachers were teaching. Thus, their criticism refers to what goes on in U.S. teacher education programs, rather than what teachers have been observed doing in their classrooms.

But, as Wallen and Travers (1963, p. 453) pointed out, "Principals commonly voice the opinion that most teachers do not teach in accordance with the pattern prescribed by teacher-training institutions, but in the pattern they observed when they were pupils ...." Similarly, Labaree (2004, pp. 129–169) elaborated on the difference between teacher education and teaching. He described what he called "the Ed School's romance with progressivism." He saw teacher education programs in the United States as strongly biased towards progressivism for many years. He wrote, "[T]he ed schools could indeed cause damage in schools if it were in the hands of an institution that was powerful enough to implement this vision, *but the ed school is too weak to do so*" [italics added] (p. 130).

Thus the Evers-Hirsch portrayal *may* be accurate if it is applied to teacher education programs, or at least what these programs advocate. But the historical and observational evidence belies their views and tells us that progressivism has been missing from U.S. classrooms for a long time.

Chapter 5
A Conception of the Content of Teaching

Passing along the content of a society's culture is what education is for. The content of teaching is derived from conceptions of the objectives of teaching – what people should know, understand, and be able to do. It is what makes education important to a society in fostering freedom and well-being, useful to individuals in achieving their economic roles, and, in a democracy, essential to the competence of citizens in making their system of government work.

We turn now to a formulation of the ways in which the content of teaching bears upon a theory of teaching. The formulation will support change in research on teaching, i.e., in the kinds of questions investigated and the kinds of data collected.

The Neglect of Content in Process–Product Research on Teaching

As research on teaching began to flourish in the 1960s, one of its most active paradigms came to be known as "process-product research" – the search for relationships between teaching *processes* (what teachers did) and the *products* of teaching (what students learned). No similarly active research movement developed to study the other main component of teaching – its *content*. Thus there was no similar level of research with what might have been called "(process ↔ content)-product research," where the ↔ symbol denotes the interaction between process and content. This chapter will explore explanations of this neglect and some of the alternative approaches to a concern with content that have developed. Those approaches will bear upon a theory of teaching, that is, an explanation of the ways in which teaching brings about students' achievement of the objectives of teaching.

The omissions and mistreatments of content in studies of teaching have been the subject of philosophical insights and also, by way of proposed remedies, the observations by educators.

N.L. Gage, *A Conception of Teaching*,
DOI: 10.1007/978-0-387-09446-5_5, © Springer Science+Business Media, LLC 2009

The Garrison–Macmillan Critiques

The neglect of content in process-product research on teaching was repeatedly criticized by two philosophers of education: Garrison and Macmillan (1984); Macmillan and Garrison (1984). But their references to content confusingly alluded sometimes to philosophers' technical term, *intentionality*, and sometimes to the everyday concept, *intention*. Thus Macmillan and Garrison (1984) wrote,

> The process-product tradition explicitly ignores the *intentions* [italics added] of teachers and learners in its investigations. ... The failure of process-product research to come to grips with the essential *intentionality* [italics added] of teaching is its greatest conceptual shortcoming. ... The unit of observation [in process-product research] is a "behavior" of the teacher precisely and technically defined so as to require little or no inference as to the teacher's intentions or to the context [intentionality?] of the behavior that might give it a broader meaning. Without some wider context, behaviors are as meaningless as physical movements of chess pieces. (pp. 18–19)

Gage and Needels (1989), lacking guidance from the philosophers, erroneously considered "intentionality" to refer to the intention, or purpose, of teachers. Their confusion was eventually resolved by a helpful entry on "intentionality" in *The Cambridge Dictionary of Philosophy* (Audi, 1995):

> intentionality, *aboutness.* Things that are about other things exhibit intentionality. Beliefs and other mental states exhibit intentionality, but so, in a derived way, do sentences and books, maps and pictures, and other representations. The adjective "intentional" in this philosophical sense is a technical term not to be confused with the more familiar sense, characterizing something done on purpose. Hopes and fears, for instance, are not things we do, not intentional acts in the latter, familiar sense, but they are intentional phenomena in the technical sense: hopes and fears are about various things. ... Phenomena with intentionality thus point outside of themselves to something else: whatever they are of or about. ... All and only mental phenomena exhibit intentionality. (p. 381)

It can be readily understood that teaching must be *about* something, namely, the content of teaching. (The "beliefs and mental states" of teachers are embraced by the category labeled "Teachers' Thought Processes" in the paradigm presented in Chap. 3.)

The problem of determining the intentionality of teachers' behaviors is generally solved, by both students and researchers, by using the process of inference: "the process of drawing a conclusion from premises or assumptions" (Audi, 1995, p. 369). When observing the verbal or nonverbal behavior of a teacher, the student or researcher usually has little difficulty in inferring the teaching's intentionality, including her beliefs and mental states. That they are able to do so results from their abundant experience in making such inferences in daily life. Indeed, daily life in any society would become chaotic if people were not typically correct in inferring intentionality from behavior. Researchers favor low-inference descriptions of behaviors not to reduce the level of inference in interpreting behavior as to content or intentionality. Rather, they do so to improve the communicability to teachers of the behaviors and intentions desired by either researchers or teacher educators.

But many researchers on teaching, in focusing on the process of teaching, have neglected what teaching was *about*, namely, its content. The successive editions of the *Handbook of Research on Teaching* (Gage, 1963; Travers, 1973; Wittrock,

1986a; Richardson, 2001) did not completely bypass content. But they considered it only briefly, as in the following passage from Shulman (1986a):

> The content and the purposes for which it is taught are the very heart of the teaching-learning processes. [B. O.] Smith (1983) put it clearly when he asserted that the "teacher interacts with the student in and through the content, and the student interacts with the teacher in the same way" (p. 491). Although the content transmitted for particular purposes has rarely been a central part of studies of teaching, it certainly deserves a place in our comprehensive map, if only to remind us of its neglect. (p. 8)

Similarly, Graber (2002) noted that:

> Content knowledge describes what a teacher understands about the subject matter .... Regrettably, investigations into this critical form of teacher knowledge are largely absent from the literature, yet dialogue continues about what forms of content knowledge are most significant for the preparation of teachers. ... (p. 495)

That the shift toward a concern with the content taught has been considered pertinent to research on teaching, was made explicit by Doyle (1995) as follows:

> The framing of pedagogical research as a question about teacher effectiveness and the grounding of this research in behavioral psychology led to an increasingly generic view of pedagogy. The language of behavioral psychology focused on overt actions, and the effectiveness question focused on teachers or teaching methods as the direct causes of outcomes. Taken together, these frameworks excluded students and *curriculum* [italics added] from the analysis. Methods became treatments, e.g., lecture versus discussion, devoid of the rich theoretical propositions about the nature of content and its acquisition that had characterized 19th-century discussions of method. *In process-product research, behaviors rather than content were measured during classroom observations. ... Questions about what was being taught to whom slipped into the background.* [italics added] (p. 491)

Emphasis on this substantive change toward taking account of content promises to improve the results of the kind of research on teaching called, until now, process-product research (see, e.g., Brophy & Good, 1986; Gage & Needels, 1989; Needels & Gage, 1991). In correlational process-product research, where teaching is studied as it occurs naturally, i.e., not manipulated through carefully targeted training of teachers by an experimenter, that relationship takes the form of a *coefficient of correlation* between process variables and product, i.e., achievement and attitude, variables. In *experimental* process-product research, manipulation of the process of teaching results from training experimental-group teachers to teach in some new ways and withholding this special training from control-group teachers. Here the relationship between process and product variables takes the form of an *effect size*, or the standardized difference in mean achievement scores between the students of the trained, or experimental, groups of teachers and the students of the untrained, or control, groups of teachers. One of several formulas for effect size (Hedges & Olkin, 1985) is shown in Fig. 5.1.

$$\text{Effect size} = \frac{\text{Mean achievement}_{\text{experimental}} \; minus \; \text{Mean achievement}_{\text{control}}}{\text{Standard Deviation of the Control Group}}$$

Fig. 5.1 A formula for the standardized size of the teachers' effect on the experimental group as compared with the teachers' effect on the control group.

The magnitude of the correlation coefficient or the effect size can tell us, in a quantitative sense, how well we "explain" the connection between teaching and achievement. (This form of explanation is analogous to the statistical concept of *amount of variance in outcome accounted for*, i.e., "explained," by the difference in teaching.)

The proposed change – the inclusion of a content-of-teaching variable – will end a deficiency that has been evident in the very label of such research until now: process-product research (PPR). In the future that kind of research should take the form of (*process ↔ content*)-*product research.* (The ↔ symbol stands for the joint consideration of process and content). The change will reflect an appreciation of a long-overdue insight, namely, that *what* is taught is at least as important for student achievement as *how* it is taught, if our purpose is to understand what it is in teaching that affects how well students perform on assessments of their achievement.

Content Variation

Content could not help explain differences in teaching effectiveness if it were a constant. A constant, by definition, lacks variance and therefore cannot explain variance in a variable, such as student achievement.

Content varies even in a single subject matter, such as arithmetic or history, in a single curriculum of that subject matter, in a single state, school district, school, or classroom. Even under research conditions, unless the researcher presents the content via some physical medium, such as a book or a computer, the same subject's content will vary from teacher to teacher. Even the same teacher will present the content so that it differs from one time to another.

This inevitability of content variability makes content difficult to control. That difficulty may explain the neglect of content in research on teaching. It may have led researchers to make the false assumption that variance in student achievement within a single subject-matter, such as fourth-grade arithmetic, could be explained solely on the basis of between-teacher differences in the *process* of teaching.

Instructional Alignment

Several authors have developed, independently of one another and with different terminologies, the same substantive advancement in research on teaching: concern with the *content* of teaching. They have introduced that concern as distinguished from concern with the process of teaching.

Although they used different terms for what they were emphasizing, all of their terms designate what S. A. Cohen (1987) called *instructional alignment: the degree of fit between (a) the content taught and (b) the content tested in assessing student achievement.* That researchers on teaching have neglected content shows up in the absence of "instructional alignment" and its various synonyms from the indexes of all four editions of the *Handbook of Research on Teaching*. S. A. Cohen's (1987) labeling instructional alignment a "magic bullet" is easy to understand; (process ↔ content)-product research should increase the ability of research on teaching to explain, to predict, and to control, i.e., improve, how well students achieve.

Approaches to Instructional Alignment

Instructional alignment was almost completely absent from the variables studied in research on teaching until now. But it has, in various forms, a substantial history in the thinking of educational research workers in other fields as against the field called research on teaching. In that history, a variety of terms and methods identified ideas similar to instructional alignment.

Content and Curricular Validity. Educational measurement has long used the terms *content validity* and *curricular validity* to designate a property of an assessment, namely, "the degree to which assessment questions are representative of the knowledge, skills, or attitudes to be learned as a result of a given teaching or course of study" (English & English, 1958, p. 575). *Or* "the extent to which a test measures a representative sample of the subject matter or behavior under investigation" (VandenBos, 2007, p. 224).

Opportunity-to-Learn. Husén (1967) reported on an international study of mathematics achievement in twelve countries. His report used the concept of *opportunity-to-learn*. It was measured by the average of teachers' ratings of each item on an achievement assessment as to the degree to which the teacher had given her students preparation for answering the assessment-item correctly.

The various participating nations were understandably sensitive to the fairness of the mathematics test used. No country wanted its students to be disadvantaged because they had not been taught any part of the knowledge or skills measured by the test.

Teachers *within* each of the 12 countries also varied in how well the content of their teaching fitted the content of the achievement test. It was reasonable to expect such variability *between* the average ratings of the teachers of different countries to be much greater than that *within* countries.

L. W. Anderson (1987) reported on a subsequent international study in nine countries. That study examined how teacher behavior was related to student achievement. Anderson also used the term *opportunity-to-learn* in offering the generalization that:

> Within countries, teachers differ greatly in what they teach relative to what was tested. As a national option teachers were asked to indicate the extent to which their students had had opportunity-to-learn the content included in each item on the posttest. On the basis of their responses, an OTL [opportunity-to-learn] index was derived for each teacher. This index represented the percentage of items on the posttest that the teachers believed their students had had an opportunity to learn during the conduct of the study ... [T]he difference between the classrooms in which the students had the most OTL and the classrooms which had the least OTL is never less that 50 percent in any country. (pp. 78–79)

The issue of "opportunity to learn" was recognized by Needels (1988) when she noted that "because the traditional [process-product] design requires that the same test items must be used for all classes, no consideration can be given to whether all content contained in the test items was actually taught by the teacher" (p. 504). She developed a process-product design, presented in Chap. 8, that permitted including only test items related to the content actually taught.

Facet Similarity. Bar-On and Perlberg (1985) used the facet theory of Guttman (1959) in developing a theory of instruction. They derived the hypothesis that the more similar two variables ("structuples") were in terms of their components ("structs"), the more highly they would correlate. Thus, Higher correlations are expected between items differing by only one struct [component of a variable] rather than between structuples [variables consisting of more than one component] differing by two structs. That is, the expected correlation between a_1b_1 and a_1b_2 is higher than the expected correlation between a_1b_1 and a_2b_2 (p. 99).

It is easy to regard the two variables (structuples) as "teaching (a_1b_1)" and "student achievement (a_1b_2)."

Measurement-Driven Instruction. Popham (1985, 1993) used the term "measurement-driven instruction" to designate teaching whose content was shaped by the test that would be used to evaluate achievement. That is, the content of the test would be developed first – either by the teacher or by the developers of a standardized test – and then teachers will be "driven" by knowledge of that test to align the content of their teaching with the content of the test.

Instructional Alignment. S. A. Cohen (1987, 1991, 1995) coined the term "instructional alignment" to designate a characteristic of teaching, namely, the degree to which the content of the teaching matched the content of the assessment used to evaluate achievement resulting from that teaching.

Measuring the Content of Instruction. A. C. Porter (2002) developed quantitative measures and graphic displays of content taught that permitted comparisons between separate bodies of content. The quantification used the following formula:

$$\text{Alignment index} = 1 - (\text{sum of } x-y/2)$$

where x denotes cell proportions in one matrix and y denotes cell proportions in another matrix.

The "matrices" here are two-dimensional tables, with the various "content topics" in the rows and various "categories of cognitive demand," or what we have called "cognitive processes," in the columns. Porter's illustrative "Content Matrix," with six content topics and five categories of cognitive demand has 6 times 5, or 30, cells. In each cell is a proportion of all the items in the teaching content or the test content. If two teachers were being compared, one would be X and the other Y. If the proportions of items in each cell were the same for the two teachers, their "alignment index" would equal 1, signifying perfect alignment between the two teachers.

Instructional Alignment Summarized

The term *instructional alignment* refers to the similarity between the content taught and the content of the assessment of achievement of the objectives of the teaching. The content taught consists of the facts, concepts, procedures, and self-understandings (metacognitions) that students should know, understand, and be able to use in various ways, as a result of the teaching. That content should conform to the curriculum that has been constructed by the persons and organizations authorized by public or private interests to define the curriculum. The curriculum often, *especially in schools*, should include some means – such as tests, observations, interviews, and portfolios (collections of things written or made by students) – for determining how well students have achieved the objectives set forth in the curriculum.

It seems obvious that the more the teaching's content resembles the assessment's content, the higher the achievement measure will be. Despite this obviousness, research on teaching has proceeded for many decades without attention to instructional alignment. The obviousness of the connection between opportunity-to-learn and student achievement may make it susceptible to disrespect by the general public. But such obviousness does not lessen its importance.

Methods of Studying Instructional Alignment

Henceforth a major concern in achieving the purposes of teaching should be instructional alignment. Research on teaching has begun to show explicit concern with that aspect of the content of teaching. Several methods have been devised to measure or manipulate instructional alignment: (a) teachers' ratings of test items as to their students' opportunity-to-learn what is required by the test items, (b) content analysis of teaching and assessment, and (c) manipulation of the alignment in an experiment.

Teachers' Ratings of Opportunity-to-Learn. A relatively early realization of the importance of instructional alignment arose in international evaluations of achievement, namely, those conducted by the International Association for the Evaluation of Educational Achievement (IEA). Teams of curriculum experts from all of the participating nations constructed the achievement tests used in that research. Those teams attempted to make the tests equally "fair" to all the nations, that is, maximally similar to the curriculum of each nation. The measure of instructional alignment was termed *opportunity-to-learn*. In the first IEA report, Husén (1967) wrote that:

> One of the factors which may influence scores on an achievement examination is whether or not the students have had an opportunity to study a particular topic or learn how to solve a particular type of problem presented by the test. If they have not had such an opportunity, they might in some cases transfer learning from related topics to produce a solution, but certainly their chance of responding correctly to the test item would be reduced. (pp. 162–163)

Opportunity-to-learn was measured by giving teachers a questionnaire on which they were to indicate, for each item of the mathematics achievement test, whether "All or most (at least 75%)," or "Some (25–75%)," or "Few or none (under 25%)" of the teacher's students had had an opportunity-to-learn how to solve that type of problem. For each teacher's class, there was thus a mean teacher rating over all test items of opportunity-to-learn and a mean score of the teacher's students on the test. The correlation between the mean rating of opportunity-to-learn and the mean mathematics test scores was calculated, across the teachers *within* each country. There was much variation between countries and between student populations within countries in the size of these within-country coefficients. Two countries had no significant coefficients, perhaps because either "opportunity-to-learn" or average achievement in the country's classrooms did not vary enough to make possible a high correlation. Two other countries had correlation coefficients higher than 0.50.

The *between*-countries correlations for these variables, that is, correlations with the country as the unit of analysis, would use measures of instructional alignment and achievement that were much more reliable because they were based on much more data concerning the "opportunity-to-learn" and "achievement" variables. And, indeed, those correlations averaged 0.64. In short, "students scored higher in countries where the tests were considered by the teachers to be more appropriate to the experience of their students" (Husén, 1967, p. 168).

In the Netherlands, Pelgrum, Eggen, and Plomp (1986) (of whose work I was informed by T. Neville Postlethwaite) used data from 15 countries that participated in a Second International Mathematics Study. As shown in Fig. 5.2, 15 nations' mean degree of opportunity-to-learn, measured across all teachers and all items, correlated substantially ($r=0.57$), with the mean achievement of the students in those countries.

But Fig. 5.2 also shows that countries with approximately similar OTLs (average measures of opportunity to learn) can have very different achievement scores. As Pelgrum et al. (1986, p. 11) noted, this means that factors other than opportunity-to-learn were also influencing student achievement.

Among these factors other than opportunity-to-learn were "teacher *and student* [italics added] ratings of test items on the question of whether the corresponding subject matter was taught" (p. 67). Pelgrum et al. validated these ratings by showing that they corresponded with the contents of the textbooks used by teachers, and also by showing that "factor analyses of the ratings reproduced the (curriculum) structure of the Dutch school system" (p. 78).

Content Analysis. Comparing analyses of the content taught and analyses of the content of the achievement test also provides a way to measure instructional alignment, illustrated by the work of W. H. Schmidt (1978). An example would be the content of a lesson on the addition of fractions and assessment items measuring ability to add fractions.

Experimental Manipulation. A third kind of evidence concerning instructional alignment was used by S. A. Cohen's student, Koczor (1984), who studied the effects of experimentally manipulated instructional alignment. An example of such manipulation is the following:

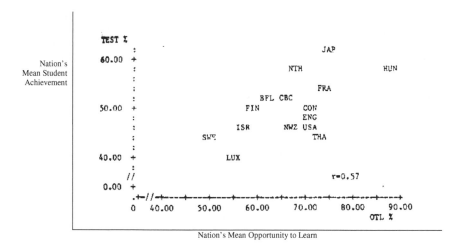

Fig. 5.2 Scattergram of mean test scores (TEST) and percentages of opportunity to learn (OTL) for 15 countries (Adapted from Pelgrum et al., 1986, p. 9.).

For example, the Roman numeral lesson taught subjects how to write the Roman numeral for arabic numerals (e. g., "Write the Roman numeral for 10."). The aligned test requested the same kind of behavior, arabic to Roman numeral. In the less aligned test, half the items reversed the request asking the subject to write the arabic numeral for the Roman numeral. In the least aligned treatment, all the items were from Roman to arabic numerals, a transfer task. (Koczor, 1984)

In this experiment, another variable was student aptitude. The results suggested that the influence of instructional alignment on achievement was greater for low-aptitude students than for high-aptitude students, perhaps because the aptitude variable was the equivalent of ability to cope with the different demands for transfer of learning imposed by the different levels of instructional alignment. This finding is, of course, an example of aptitude-treatment interaction (Cronbach & Snow, 1977), in that the effect of the treatments (different degrees of instructional alignment) was different for students differing in aptitude.

Measurement-Driven Instruction. A fourth aspect of instructional alignment is measurement-driven instruction (MDI) (Popham, 1985, 1993). When such "driven-ness" occurs, i.e., when the alignment of the teaching content with the achievement-assessment content, *designed in advance*, is very high, achievement should also be high. But if the teaching was not "driven" by what the subsequent assessment would measure, the alignment of the teaching with the assessment would be low, and achievement would be low.

The proper alignment of the teaching content with the assessment content was argued in the classic statement by Tyler (1951) on the desirable relationship between teaching and testing – a relationship that might be labeled "instruction-driven-measurement" (IDM):

Viewed in this way, instruction involves several steps. The first of these is to decide what ends to seek, that is, what objectives to aim at or, stated more precisely, what changes in students' behavior to try to bring about. The second step is to determine what content and learning experiences can be used that are likely to attain these ends, these changes in student behavior. The third step is to determine an effective organization of these learning experiences such that their cumulative effect will be such as to bring about the desired changes in an efficient fashion. *Finally, the fourth step is to appraise the effects of the learning experiences to find out in what way they have been effective and in what respects they have not produced the results desired.* [italics added] (p. 48)

High-Stakes Assessment. The term *high-stakes assessment* refers to assessment that has extremely important consequences, such as a student's graduation or the evaluation of a teacher's competence, High-stakes assessment typically uses standardized rather than teacher-made assessments. These assessment tools are constructed for making comparisons between students – and also between their teachers. They are constructed by teams of curriculum and measurement experts rather than by teachers, who are – in some conceptions of teaching – permitted or expected to put their personal "stamp," – their own emphases and de-emphases – on the curriculum. High-stakes testing, according to Nichols and Berliner (2007), typically falls very short of achieving its purposes. It does not motivate teachers and their students to improve their performance. Instead, it tends to bear out "Campbell's Law," (named after Donald T. Campbell, the psychologist who first called attention to it): the tendency to corruption resulting from the use of any quantitative measures, such as test scores, as the basis for social decision-making. High-stakes testing leads to cheating by administrators, teachers, and students.

Controversy over MDI has resulted from the conviction of some critics (e.g., Bracey, 1987a, b) that it will lead to "teaching to the test." Tests are almost always intended to measure a *sample* rather than to measure the "universe," (the whole array), of outcomes at which the teaching should aim. In social studies, for example, only a sample of the amendments to the U.S. Constitution would be the subject of test questions. In mathematics, only a sample of the applications of quadratic equations would be the subject of test questions.

Measurement-driven-instruction would reward the teacher's concentration on just the sample, *known by the teacher in advance*, instead of the educationally more valuable universe of the knowledge and skills to be taught. The resulting measures of achievement will be spurious in the same sense that a measure of the safety of a sample of water treated with anti-bacterial agents would be spurious if it was considered to measure the safety of the untreated total supply of water.

The best policy is twofold: (1) "curriculum-driven-teaching," and (2) "curriculum-driven-assessment." For *teacher-made* assessments, such a policy puts the teaching for the teacher's class in the hands of the teacher, who may have had to shape the curriculum to adjust to characteristics of her students, her school, and her community. For *standardized* tests – constructed not by the teacher but by a team of experts on the curriculum and on test-construction and intended for use in many classes – the teaching should aim at the teacher's conception of the best sample from the population of objectives. For teacher-made assessments, constructed and used by a teacher for her own class, it would be necessary to measure instructional alignment by some means such as content analysis of both the teaching and the assessing. With the teacher as the sole determiner of both, it is reasonable to expect her instructional alignment to be high. But it could not be judged in comparison with that of other teachers, who teach and test in their own distinctive ways, except in terms of a percentage of the "items" taught that are similar to the items tested.

Standardized tests are useful for comparing the achievement of many classes for two purposes: research on teaching and high-stakes assessment. For research on teaching, the scores are used to compare teaching practices as to their effect on student achievement. In high-stakes testing, test scores affect important matters such as student promotion or the evaluation of a teacher. Research on teaching uses the same test, constructed or selected by the researcher, for all the teachers studied so that the researcher can compare teachers' classes with one another, and the correlates of the achievement of the teachers' classes can be determined. High-stakes assessment leaves the content of the assessment in the hands of measurement experts – employed by states, counties, or cities – who construct standardized assessments for assessing the achievement of all the comparable classes of a specified grade-level. Here again the individual teacher does not determine the instructional alignment; it is left, usually to an unknown degree, to the unforeseeable similarity between the teacher's content and the content of the high-stakes assessment. If the teachers learn in detail the content of the high-stakes assessment, they will understandably try to increase their instructional alignment, i.e., to engage in "teaching to the test."

To leave the curriculum entirely in the hands of the individual teacher, as instruction-driven measurement would have it, is to abandon the idea of a fairly uniform curriculum for the students in all of the fifty states of the U.S. Such a solution makes the curriculum determine, i.e., serve as the arbiter of, the content of both the teaching and the testing.

Does high-stakes testing influence the curriculum taught by individual teachers?

Some writers have proposed that high-stakes assessments should emphasize critical thinking, a kind of achievement that has typically been neglected in both standardized and teacher-made assessments. Teaching to such assessments would be desirable as a way of reducing the emphasis of both teacher-made and standardized assessments on memorized knowledge and increasing the emphasis on critical thinking.

Such desirable new emphases in teacher-made and standardized assessments would include the kinds of higher cognitive processes carefully defined in the *Taxonomy of Learning, Teaching, and Assessing* (Anderson, 2001), namely, *understanding, applying, analyzing, evaluating, and creating*. Teaching toward such assessments would, many writers hold, improve the content of teaching. But, as we note in Chap. 9, Hirsch (1996) has argued against the desirability of teaching for critical thinking on the grounds that critical thinking requires large stores of relevant knowledge attainable only via large expenditures of time.

How can instructional alignment be achieved without teaching-to-the-test of an obviously unethical kind? We can put the issue as a question of *transfer demand*, or the amount of transfer from an item that was taught to what is demanded by an item in the subsequent achievement-test. Transfer is "change in ability to perform a given act as a direct consequence of having performed another act relevant or related to it" (English & English, 1958, p. 562). Transfer demand would be at a minimum if the teacher taught directly to the test, providing training in answering the identical questions in both the teaching and the testing. But if the test content requires some transfer of what was taught to what the test calls for, teaching-to-the-test becomes increasingly desirable. Then it is equivalent to teaching for positive transfer, that is, teaching content and skills that will facilitate answering questions, performing tasks, or solving problems presenting transfer demand from those that were taught.

Thus teaching-to-the-test is desirable insofar as the "assessment" imposes a reasonable transfer demand. Teaching and assessing for critical thinking would require a reasonable transfer demand, inasmuch as critical thinking by definition requires argumentation on issues not previously encountered and thus a suitable amount of transfer demand. So we can resolve the apparent conflict between desirable instructional alignment and undesirable teaching-to-the-test insofar as the assessment imposes a transfer demand that is reasonable in the light of the student's maturity and the teacher's educational objectives.

Categorizations of Content

What students are taught is customarily put into the well-known categories called school subjects. In elementary schools, the subjects customarily include reading, writing, arithmetic, social studies (history, geography), science, art, music, and physical education. In secondary schools, the subjects include English (composition and literature), mathematics (algebra, geometry, trigonometry, and, in recent decades, calculus), social studies (history, civics), foreign languages, and science (biology, chemistry, physics).

For each of these subjects, there are subject-matter, or curriculum, specialists who have created bodies of scholarship and research (e.g., Brophy, 2001; Jackson, 1992) on the content that is best included (and excluded). Various journals serve their authors. Issues arise, controversies flourish, and schisms form.

The treatment of content categorization in this chapter will not draw upon the work of subject-matter specialists. Rather it will deal with the ideas of educational psychologists who have focused on the general phenomena of learning and teaching – phenomena that occur in the form and of problems and their solutions, that cut across subject matters. These psychologists have developed categorizations of content – categorizations to which we now turn.

Taxonomies

Several major taxonomies – systems of classification – of content have been developed. Each attempts a classification of concepts relevant to content. We shall deal only with those that bear upon the *cognitive* – as distinguished from *social-emotional* and *psychomotor* – educational objectives.

Bloom's Taxonomy

The first of these was the famous *Taxonomy of educational objectives. The classification of educational goals. Handbook 1: Cognitive domain* (Bloom, Engelhart, Furst, Hill, & Krathwohl, 1956). It was developed to serve examiners – developers of achievement tests – who needed carefully developed definitions and categorizations of the kinds of achievement that their examinations – consisting of test items, questions, and problems – would seek to evaluate. Its categories are summarized in Table 5.1.

Table 5.1 Categories of the bloom taxonomy of educational objectives in the cognitive domain

Knowledge
1.00 Knowledge
1.10 Knowledge of specifics
1.11 Knowledge of terminology
1.12 Knowledge of specific facts
1.20 Knowledge of ways and means of dealing with specifics
1.21 Knowledge of conventions
1.22 Knowledge of trends and sequences
1.23 Knowledge of classifications and categories
1.24 Knowledge of criteria
1.25 Knowledge of methodology
1.30 Knowledge of the universals and abstractions in a field
1.31 Knowledge of principles and generalizations
1.32 Knowledge of theories and structures
Intellectual abilities and skills
2.0 Comprehension
2.10 Translation
2.11 Comprehension
2.12 Extrapolation
3.00 Application
4.00 Analysis
4.10 Analysis of elements
4.20 Analysis of relationships
4.30 Analysis of organizational principles
5.00 Synthesis
5.10 Production of a unique communication
5.20 Production of a plan, or a proposed set of operations
5.30 Derivation of a set of abstract relations
6.00 Evaluation
6.10 Judgments in terms of internal evidence
6.20 Judgments in terms of external criteria

Source: Bloom et al. (1956)

The Anderson–Krathwohl Taxonomy for Learning, Teaching, and Assessment

This taxonomy, developed by Anderson et al. (2001), aimed to serve not only the purposes of assessing but also those of learning and teaching. Their work led them to set up four categories of Knowledge: *Factual, Conceptual, Procedural,* and *Metacognitive. (Their five categories of Cognitive Process are treated in Chap. 8).* Table 5.2 gives the subcategories of the four categories of Knowledge.

Table 5.2   The knowledge: major types and subtypes

A. Factual knowledge
   Aa. Knowledge of terminology
   Ab. Knowledge of specific details and elements
B. Conceptual knowledge
   Ba. Knowledge of classifications and categories
   Bb. Knowledge of principles and generalizations
   Bc. Knowledge of theories, models, and structures
C. Procedural knowledge
   Ca. Knowledge of subject-specific skills and algorithms
   Cb. Knowledge of subject-specific techniques and methods
   Cc. Knowledge of criteria for determining when to use appropriate procedures
D. Metacognitive knowledge
   Da. Strategic knowledge
   Db. Knowledge about cognitive tasks, including appropriate contextual and conditional knowledge
   Dc. Self-knowledge

Source: Anderson et al. (2001)

Types of Knowledge

The teacher's structuring can state factual knowledge, conceptual knowledge, procedural knowledge, or metacognitive knowledge.

Factual Knowledge. This category consists of information about what is true or false. It takes such forms as *propositions* (Bill Clinton has been President; An even number is divisible by two), *images* (Alaska is northwest of Canada; Giraffes are taller than elephants), and *linear orderings* (The glider was invented before the airplane; Vowels are usually listed as *a-e-i-o-u-* and sometimes *y).*

Conceptual Knowledge. This category includes general ideas (honesty, mass, democracy), principles (mass equals force times acceleration), distinctions (rights differ from obligations).

Procedural Knowledge. This category comprises ways of doing things, such as *behavioral* (Walk carefully; Start sentences with capital letters) and *mental* (Think before you respond; Summarize after you've described; Review after you've studied).

Metacognitive Knowledge. This category consists of thinking about one's own cognitions. This takes such forms as asking yourself whether you know or understand, whether you should reread or review, whether you're allowing enough time for carrying out a task, whether you've explained something clearly enough.

The role of content in a theory of teaching is to denote *what* is being taught, just as the role of process in the theory is to denote *how* it is being taught. We turn now, in Chap. 7, to salient characteristics of the students to whom teaching is addressed – their cognitive capability and their motivation.

Chapter 6
Conceptions of Students' Cognitive Capabilities and Motivation

The process and content of teaching must be appropriate to two major categories of factors – the students' cognitive capabilities and the students' motivation. Concern with students' cognitive capabilities and motivation should take place early in any teacher's thinking about the process and content of teaching.

Two Components of Cognitive Capabilities

Cognitive capabilities are two-sided. They comprise (a) the student's intelligence and (b) the student's prior knowledge of what is being taught. The intelligence component applies to content in *general*. Prior knowledge consists of what students know, before being taught, about some *specific* kind of content.

Intelligence

We also call intelligence by such other names as intelligence quotient (IQ), general intellectual capability, scholastic aptitude, cognitive ability (Carroll, 1993), and multiple intelligences (Gardner, 1983). It has a general form – general intelligence – that applies to all kinds of cognitive tasks, and a number of more specific forms that apply to more specific kinds of intellectual tasks, such as verbal, mathematical, and spatial.

The special abilities have been identified in two ways: (a) factor analysis – "a mathematical procedure for reducing a set of intercorrelations to a small number of descriptive or explanatory concepts" (VandenBos, 2007). This statistical method was used on a large scale by Carroll (1993) for determining the categories, or "factors," into which various kinds of cognitive abilities fall. Such categorization was also performed by Gardner (1983) using a "multiple-intelligences" approach described below.

Cognitive capabilities, called "mental age" by the original developers of intelligence tests, make up the numerator in the traditional definition of the intelligence

N.L. Gage, *A Conception of Teaching*,
DOI: 10.1007/978-0-387-09446-5_6, © Springer Science+Business Media, LLC 2009

quotient, or IQ, which equals *Mental age* divided by *Chronological age*. Mental age is "the level of development in intelligence, expressed as equivalent to the life [chronological] age at which the average child attains that level" (English & English, 1958, p. 18).

Factor Analysis. Factor analysis is the statistical analysis of the intercorrelations between scores on a considerable number of tests, all of which, for the purposes of this chapter, measure some aspect of intelligence. Its purpose, as considered here, is to search for factors, or clusters of tests, that measure the same kind of mental ability. It yields special and general factors. The special factors appear when factor analysis identifies clusters of tests that measure to a high degree the same underlying, abstract kind of special ability, such as verbal ability, mathematical ability, and spatial ability. The tests within a factor correlate more highly with other tests in the same factor than they do with tests in other factors. The basis for the concept of *general intelligence,* dubbed *"g"* by its discoverer (Spearman, 1904), rests on the fact that all the special abilities have always been found to correlate positively with one another. That is, people who score at a given level on one ability, say verbal, *tend to* score at *approximately* the same level on other abilities, say, mathematical and spatial. Those positive correlations taken together betoken an underlying *general ability* that can work well on all of the kinds of cognitive content.

Carroll's (1993) integration of the findings of factor analyses yielded the kinds of special abilities listed in Table 6.1.

Multiple Intelligences. Special abilities were also identified by Gardner (1983) from a variety of kinds of evidence: literary accounts, neurological evidence, descriptions of genius and deficiency, and anthropological reports. His findings, termed *multiple intelligences*, resemble and also differ from Carroll's, as is shown in Table 6.1.

Prior Knowledge

The other part of cognitive capabilities is the student's *prior knowledge* of the content being taught. Students often vary in prior knowledge according to their interests, self-education, previous schooling, and random experiences. Whatever the

Table 6.1 A comparison of Carroll's and Gardner's cognitive-ability categories

Carroll's cognitive-ability factors	Gardner's multiple intelligences
Crystallized intelligence	Linguistic
Auditory perception ability	Musical
Fluid intelligence	Logical-mathematical
Visual perception	Spatial
[No counterpart]	Bodily-kinesthetic
Knowledge-of-behavioral-content	Interpersonal

Source: Carroll (1993, p. 461)

content, or subject matter, of the teaching may be, it is reasonable to assume that students will differ, sometimes substantially, in their prior knowledge of that content. Those differences should influence the effectiveness of the process and content of the teaching aimed at those students.

The Russian psychologist Vygotsky (1978) recognized the importance of prior knowledge in his concept of "zone of proximal development" (ZPD), formulated in the 1920s. The ZPD consists of the difference between the knowledge, or problem-solving ability, of the student *before* receiving instruction and the student's capabilities *after* receiving the guidance and assistance of a teacher. Ausubel (1968, p. iv) recognized the reasonableness of the assumption that prior knowledge is important when he stated that "If I had to reduce all of educational psychology to just one principle, I would say this: The most important single factor influencing learning is what the learner already knows. Ascertain this and teach him accordingly." Or, as R. M. Gagné (1985) made the same point,

> One set of factors contributing to learning is the capabilities that already exist in the individual before any particular new learning begins. The child who is learning to tie shoelaces does not begin this learning "from scratch" but already knows how to hold the laces, how to loop one over the other, how to tighten the loop, and so on. (p. 16)

Prior knowledge consists, for any subject matter, of the same types of knowledge (factual, conceptual, procedural, and metacognitive) and the same cognitive processes (remember, understand, apply, analyze, evaluate, and create) that appear in *A Taxonomy for Teaching, Learning, and Assessing* (Anderson & Krathwohl, 2001). Students' prior knowledge almost always varies whenever students are assessed as to their *prior* achievement of the objectives of the teaching. Typically, students' scores on a pretest of achievement correlate positively with their scores on a posttest. That is, students who had more of the types of knowledge and cognitive-process capabilities before the teaching tend to score higher after the teaching on an assessment of that content knowledge and cognitive-process capability.

Adjusting Teaching to Students' Cognitive Capabilities

Teaching should be appropriate for students' cognitive capabilities as teachers seek to carry out the tasks imposed by the teaching. Traditionally, curriculum developers for any of the various grade levels and subject matters try to adjust the content to the estimated cognitive capabilities of the anticipated students of a specific grade level. The authors of textbooks in a specific subject matter for a specific grade level must also estimate the cognitive capabilities of anticipated students.

But curriculum developers and textbook authors can adjust their product to only a relatively narrow range of cognitive capabilities. That range is narrower than the range of cognitive capabilities usually encountered by a teacher among the students in her class. For example, a fifth-grade teacher may have students whose mental ages range a full year or more above and below the average of her students.

Early Cognitive Capabilities

By the time 5- or 6-year-olds enter school, they have developed well past the infant and early childhood stages of cognitive capabilities described by Piaget and Inhelder (1973) and others. R. L. Atkinson, R. C. Atkinson, E. E. Bem, & S. Nolen-Hoeksema (2000, p. 77) summarized the abilities of typical kindergarten and first-grade children as follows. Such children can typically

- differentiate themselves from objects,
- recognize themselves as agents of action and act intentionally,
- realize that things continue to exist when no longer visible or audible,
- use language to represent images and words,
- differentiate themselves from others,
- take the viewpoint of others, and
- classify objects by one or more features and arrange them in some order.

Then, between the ages of about seven to twelve, they become able to

- think logically about objects and events
- realize that objects stay the same in number and weight even when they differ in position, shape, or color,
- classify objects according to several features and order them in series on a single dimension, such as size,
- think logically about abstract propositions and test hypotheses systematically,
- become concerned with the hypothetical, the future, and ideological problems.

What these research findings on intelligence mean is that the average student is well able to cope with the elementary- and secondary-school curricula that have become well established around the world. But the students differ in the speed and facility with which they can learn what is usually taught in those grades. Students eventually succeed if they are not too far below the average level of cognitive capabilities for their age.

Cognitive Capabilities and Teaching Processes

Here we develop connections between (a) the various levels of cognitive capabilities that occur in grades 1 to 12 or at the college level and (b) the different ways of teaching that become appropriate for students at different levels of cognitive capabilities relative to the average level for their grade level.

The psychology of individual differences, which is the study of the psychological ways in which individuals and groups differ from one another, was early seen as a source of knowledge that would be useful for teachers in coping with the fact that each of their students differed in cognitive capabilities from their other students. We turn now to attempts to develop ways of dealing with this unavoidable characteristic of classrooms-full of students.

Applying Aptitude–Treatment Interactions. One method originally proposed for use by teachers in coping with individual differences among their students in cognitive capabilities was the use of findings from research on aptitude-treatment-interactions (ATIs), discussed in Chap. 5. Such interactions would enable teachers to select "treatments," or teaching processes and contents, according to how they "interacted" with students' cognitive capabilities. An ATI would mean that a given kind of teaching would yield different results for students who differed on some "aptitude," or characteristic, such as cognitive capabilities. Such interactions would mean that students at one level of cognitive capabilities would be better taught by one kind of teaching, while students at another level would be better taught by a different kind. For example, students at higher levels of cognitive capability would be taught the Pythagorean theorem with words and numbers, but students at lower levels would be taught with wooden squares and a right triangle drawn on a board.

This ATI proposal faltered on the low or non-existent replicability of findings from research on meta-analyses (e.g., Kulik, 1981). The low replicability meant that findings were inconsistent from one investigation to another and hence untrustworthy.

Also ATIs confronted teachers with ideas that were difficult to put into practice. It called upon teachers to use possibly radically different approaches to teaching for students within the same class. It called upon teachers to divide their students into two or more sharply differentiated groups. Coupled with the undependability of ATIs, these flaws brought the ATI approach to an end.

Simplifying

One effective way of teaching for different levels of cognitive capabilities within a single grade level is *simplification*. It can take various forms:

Reduce the Number and Length of the Items Encountered. For example, the teacher can reduce (a) the number of lines of a poem to be memorized; (b) the number of chemical atomic weights to be remembered; (c) the number of words in a single sentence of an explanation; (d) the number of sentences spoken by the teacher without pausing or allowing interruptions; (e) the number of sentences that refer to a given topic; (f) the length of reading assignments; (g) the number of problems assigned for homework, (h) the duration of uninterrupted teacher talk; (i) the time gap between the content of recent teacher discourse and the occurrence of teachers' questions on that discourse.

Use Simpler and More Familiar Words. Examples: "hide" rather than "conceal;" "make unclear" rather than "obfuscate."

Provide Mnemonic Devices. Example a: Mnemonic device HOMES for the names of the Great Lakes (Huron, Ontario, Michigan, Erie, Superior). *Example b:* Mnemonic Device: MVEMJSUNP = My Very Earnest Mother Just Served Us Nine Pickles. For the names of the planets, in order of their distance from the sun: Mercury, Venus, Earth, Mars, Jupiter, Saturn, Uranus, Neptune, Pluto. *Example c*: Mnemonic Device: ROY G. BIV (a made-up name) for the colors of the rainbow, in order: Red, Orange, Yellow, Green, Blue, Indigo, Violet. – OR – Richard Of York Gave Battle In Vain.

Provide Algorithms, or Sets of Rules, for Solving Problems. Example a: A set of steps for multiplying by two or more digits, whereby the product of the multiplicand (the number to be multiplied), e.g., 250, and the first digit (2) of the multiplier, e.g., 72, is written on the first line and the product of the multiplicand and the second digit of the multiplier is written on the second line and *indented one space to the left*, and so on for the multiplier's remaining 1750

$$
\begin{array}{r}
250 \\
\times 72 \\
\hline
500 \\
1750 \\
\hline
18000
\end{array}
$$

Reducing Cognitive Load

The degree to which a learning situation requires the student to exert cognitive effort has been termed *cognitive load* by Sweller (1999). As the term suggests, the concept is concerned with the "consequences of difficulty in dealing with the elements of a task," and "difficulty in holding and processing many elements in a limited working memory" (Sweller, 1999, p. 36).

Sweller (1999) elaborated on factors affecting cognitive load, namely, (a) interactivity, (b) split attention, (c) worked examples, and (d) redundancy. He also suggested ways of coping with differences in cognitive capabilities by reducing cognitive load in ways set forth in Chap. 8.

A Clinical Approach

The term "clinical" refers to the diagnosis and treatment of psychological problems, such as the adjustment of teaching to the needs of individual students. (It also refers to medical or psychiatric processes.) The teacher, in using these ways of meeting the needs of their less cognitively capable students, must momentarily treat them in ways that are different from those suitable for the other students in her class. Cronbach (1967) captured the way in which teachers typically – on the evidence of one's own memories of classroom teachers' processes – cope with the inevitable differences among their students:

> The teacher adapts instructional method to the individual on both the micro (small) and macro (large) levels. He barely acknowledges the comment one pupil makes in class discussion, and stops to praise a lesser contribution from another who (he [the teacher] thinks)

needs special encouragement. He turns away from one pupil who asks for help – "You can find the answer by yourself if you keep at it" – and walks the length of the classroom to offer help to another, because he has decided to encourage the independence of the former pupil and to minimize the frustration of the latter. On the larger scale, he not only allows options for a term paper, but may custom tailor a project for the student with special abilities or limitations.

The significant thing about these adaptations is their informality [italics added]. The teacher picks up some cues from the pupil's test record and his daily work, and other cues from rather casual observation of his social interactions. The teacher forms an impression of the pupil from the cues, usually without an explicit chain of reasoning. He proceeds on the basis of the impression to alter the instruction; the adaptation too is intuitive, without any explicit theory. No doubt the decisions tend to be beneficial, but there is reason to think that intuitive adaptations of this kind will be inefficient and occasionally may be harmful. (pp. 28–29)

At this point Cronbach introduced a warning based on his research on clinical processes. He warns against teachers' "overdifferentiation." He cites his finding (Cronbach, 1955, pp. 182–183) that counselors expect too much from high scorers and too little from low scorers on tests. The implication is that *teachers should avoid extremes in their attempts to adapt their teaching to differences in the cognitive capabilities of their students.* They will make fewer and less extreme errors when they lean toward treating all their students as close to the average.

Tutoring

Tutoring consists of the teacher's working with a single student or a small number of students who impress the teacher as needing special attention. That need can arise because the student is finding the content as presented too difficult or too easy as compared with the majority of students in the class.

For students having difficulty, the tutoring can be organized according to the same structuring, soliciting, responding, and reacting that occur in regular CDR teaching. Structuring can provide simpler and shorter explanations of content, using the simplification devices described above. Soliciting can provide easier-to-answer-correctly questions. The student's responding can be strongly encouraged, and the teacher's reaction can be warm and patient. The teacher may insure that students have adequate time-on-task, encourage students to try harder, and assure students of her expectations of their eventual success.

Teaching with Multiple Intelligences

How should the teaching process be formulated in view of the evidence of multiple intelligences? Must radically different processes be used for students known to differ in the strength of the various intelligences? Or can the Conventional-Direct-Recitation (CDR) process, delineated in Chap. 4, serve and be adapted to students' multiple intelligences?

Students' cognitive capabilities are still often considered to be unidimensional, that is, to fall higher or lower on a single measure, such as IQ or prior knowledge. So teachers could then use the kind of clinical judgment sketched above in adapting their teaching to individual differences among students on that single dimension.

In view of the identification of multiple dimensions, one approach is to revise the teaching process radically (in the sense of going to the root of the problem) so as to exploit and enhance one or more of the multiple intelligences. Then the teaching process calls forth the kind of student intelligence required by the content being taught.

Lazear (1992) developed such a radical approach. He identified four stages in such teaching:

(a) awakening the intelligences; (b) amplifying the intelligences; (c) teaching with the intelligences; and (d) transferring the intelligences.

Awakening the Intelligences. The teacher here can activate linguistic intelligence by calling for a written composition. Or she can call attention to logical-mathematical intelligence by asking for the proof of the Pythagorean theorem.

Amplifying the Intelligences. The teacher can strengthen musical intelligence by asking students to practice the scale. Or she can improve visual-spatial intelligence by asking students to draw, from memory, a map of the state they live in.

Teaching with the Intelligences. The teacher can improve students' bodily-kinesthetic intelligence through helping students understand the concept of *balance* by using a see-saw and varying their distances from the point on which the see-saw rests. Or she can enhance their interpersonal intelligence by having small groups of students discuss the pros and cons of capital punishment.

Transferring the intelligences. Here the teacher's task is to help the students understand the applicability of a teacher's structuring can adjust to her students' spatial intelligence by using visual aids ranging from a chalkboard to slides and movies projected onto a screen. Because students may differ in their ability to understand such visual aids, the teacher can modify the aids by making them simpler or more complex according to students' needs. Particularly in geometry and visual arts—such as drawing, painting, mechanical drawing, and lettering—teachers can vary such factors as the size, complexity, perspective, color, and detail in adjusting visual stimuli to students' spatial intelligence.

Musical Intelligence. Students are likely to differ greatly in musical intelligence not only because of innate factors that make them more or less musically intelligent for no observable reason. They also differ because of environmental factors such as their parents' and friends' musical interests. The elementary school teacher's responsibility is to give her students an introduction to music that will enhance their pleasure and alertness to the possibility that some of her students should get more than such an introduction, namely, an opportunity to take lessons in playing a musical instrument or in singing.

Because her structuring aims at the class as a whole, it is typically concerned with musical appreciation, not performance. As such, her structuring becomes a verbal activity for providing knowledge about music of the kinds identified in the

Anderson-Krathwohl *Taxonomy* described in Chap. 5: factual knowledge, conceptual knowledge, procedural knowledge, and metacognitive knowledge.

Soliciting for Multiple Intelligences

The kinds of questions asked by the teacher after her structuring should also be appropriate to the students' linguistic, logical-mathematical, spatial, and musical intelligences. Adjustments of soliciting for *linguistic intelligence* result from varying the same factors as those identified in structuring, namely, sentence structure, transition between paragraphs, complexity, demand for prior knowledge, and duration. Questions can be adjusted for *logical-mathematical intelligence* by varying the factual, conceptual, procedural, and metacognitive knowledges asked for. Questions can take *spatial intelligence* into account by varying the size, complexity, perspective, color, and detail to which the questions refer. Differences in *musical intelligence* can call upon the teacher to vary her questions about the musical compositions she uses in fostering music appreciation. For students qualified for instruction in some form of musical performance, she can ask for performances appropriate to the student's level of skill.

Responding and Reacting for Multiple Intelligences

We combine the treatments of responding and reacting to recognize their close psychological and temporal relationship in the practice of teaching. That close relationship makes it appropriate to discuss them together. Sirotnik (1983, p. 20) reported that, in The Study of Schools, "[T]he most frequently occurring single interaction is one of students responding to the teacher.... This occurs roughly 15 percent and 10 percent of the time at the elementary and secondary levels, respectively."

After hearing one of the kinds of questions described above, students will respond in ways that throw light on their multiple intelligences. Their responses will guide the teacher toward improved understanding of their students' capability in the kind of intelligence tapped by the question. Then the teacher will take steps to remedy any shortcomings revealed by the student's response.

A Conception of Student Motivation

Students' cognitive capabilities may be regarded as tools that students can use for the purpose of learning. The concept of motivation refers to the degree to which students use those tools. Student achievement depends on students' motivation as well as it does on their cognitive capabilities. Unmotivated bright students may learn less than well-motivated students who have less cognitive capability. The teacher's task includes maximizing the degree to which students use their cognitive capabilities for the purpose of achieving the objectives set forth in the content of teaching.

That students differ in their motivation is easily and widely recognized. Some students find school work interesting and challenging, They enjoy hearing and reading about new ideas, and they find their classroom sessions enjoyable. Other students have the opposite reactions to their classroom experiences, which they find dull, frustrating, and irrelevant to their interests.

Teachers learn to hope for the first kind of students and to dread the second kind. The first kind makes teachers feel successful, and the second kind frustrates teachers. The influences that affect student motivation are of several kinds. One kind consists of the family background. Better educated parents more often influence their children to value education more and to take advantage of opportunities to learn. Less well educated parents pass on to their children less respect for what schools offer. Similar influences come from a student's siblings and friends. Harris (1998) brought together much evidence that children are influenced more by their peers than by their parents.

What makes these differences occur? One place to look for the causes of these differences is the homes of the students. Some students' parents create home environments that foster high respect for what schools offer, and that respect makes their children more highly motivated to do well at school. Similar influences come from the student's peers, who pass their attitudes toward schooling on to their friends.

We shall focus here on the approaches that teachers can take to motivate their students. The approaches fall into the same two major categories of approaches that characterized the science of psychology during the twentieth century: behavioristic approaches and cognitive approaches.

Behavioristic Approaches

Behaviorism is an approach to psychology that focuses on observable, e.g., visible and audible, phenomena. As applied to motivation, it focuses on *operant conditioning,* namely, the changing of behavior (resulting from stimuli of unknown origin) by means of its subsequent positive or negative reinforcement. Positive reinforcement is a post-behavioral event that increases the probability of the behavior that preceded the positive reinforcement. Negative reinforcement is the withdrawal, or cessation, of an event that tends to reduce the probability of the occurrence of the preceding behavior. Thus both positive and negative reinforcement tend to increase the probability of the prior behavior.

In teaching, the teacher's praise is a positive reinforcer in that its occurrence tends to strengthen, or increase the likelihood of, the prior event, e.g., the student behavior that preceded the praise. Similarly, in teaching, the teacher's frown is a negative reinforcer whose ***non-occurrence*** strengthens the likelihood of the student's making a correct response.

Contracts. Thus students can be motivated to try to learn by either positive or negative reinforcement, and teachers presumably have for millennia been aware of this effect of their behavior (e.g., praise or frowns). But teachers can make such changes in student behavior more likely by entering into *contracts* with students. For example, student volunteering can be increased by a contract whereby the student's volunteering at least once an hour for several days will be rewarded with something the student values, such as time with a new computer. And the terms of the contract can be gradually changed until the student volunteers much more often without the extraneous reward, or reinforcement.

Token Economies. Arrangements, whereby desirable student behaviors are reinforced with "tokens" that the students can exchange for things they want are called "token economies" (McLaughlin & Williams, 1988). The tokens may be mere cumulative points, chips, stars, ratings, or check marks. The target behaviors to be decreased (e.g., talking out of turn, off-task behavior, hyperactivity) or increased (e.g., attention to school subjects, assignment completion, accuracy of performance) have been found controllable through token economies. The teacher makes sure that the rules for gaining or losing tokens are understood by her students. The things for which tokens can be exchanged take such forms as edibles, toys, trinkets, recess duration, and free time.

The effectiveness of token economies in bringing about the desired changes in student behavior has been well-documented (McLaughlin & Williams, 1988, p. 471). It represents a triumph of behavioral methods for manipulating motivation.

Cognitive Approaches

> During the last third of the twentieth century, psychologists overcame their hesitancy about theorizing with variables that were not directly observable behaviors. They became willing to work with cognitions: "all forms of knowing and awareness, such as perceiving, conceiving, remembering, reasoning, judging, imagining, and problem solving" (VandenBos, 2007, p. 187). When it came to understanding motivation, they saw a motivated person "as someone with cognitions or beliefs that lead to constructive achievement behavior, such as exerting effort or persisting in the face of difficulty" (Stipek, 2002, p. 10).

Teachers can affect cognitive aspects of motivation through what they say to students about their own expectations of students. The teacher's genuine expectation that a student can succeed at a task will motivate the student to exert effort to meet the teacher's expectations. Teachers can influence their students' perception of learning tasks: the task's *relevance* to students' values, its value for *understanding* the logical aspects of reading, its *usefulness* in using arithmetic and other aspects of mathematics in occupational and recreational activities.

Increasing a student's level of success with school tasks will lower the probability of the student's becoming unmotivated. Here the teacher's sensitivity to every student's successes and failures in meeting the demands of the curriculum can guide the teacher's treatment of the student. The student's failure to understand something should become a signal calling forth the teacher's increased effort to bring about the understanding. In short, the teacher's various cognitions should guide the teacher's effort to help every student succeed.

Chapter 7
A Conception of Classroom Management

Classroom management is a prerequisite of the process, content, and student cognitive capabilities and student motivation components treated in Chaps. 4-6. It can be distinguished from teaching because it does not deal *directly* with the process and content of teaching. As Johnson and Brooks (1979) put it,

> [T]he function of classroom management can be distinguished conceptually from the teacher's primary function, instruction, however intimately the two may be related in practice....[E]ach of the two functions can occur without the other, since instruction does not always involve a group of learners in a classroom, and classrooms need to be managed whether or not instruction is taking place. (p. 1)

The distinction between teaching and classroom management was also identified by Brophy and Good (1986) when they noted that

> [S]o many findings [of process-product research] were derived from naturalistic situations where teachers varied drastically in their allocation of time to academic activities and in their classroom organization and management skills. The differences in student opportunity to learn created by these differences in time allocation and classroom management probably overwhelmed, and thus masked, the effect of whatever differences occurred in quality of instruction. (p. 367)

The publication of a *Handbook of Classroom Management* (Evertson & Weinstein, 2006) implied further recognition of the distinction between teaching and classroom management.

Poverty

Before we consider classroom management at the classroom level, we must not shy away from a crucial aspect of the *context* of teaching – one that shapes the problem of teaching in areas of society where classroom management is most difficult. That aspect is poverty – the poverty of students, of their families, and of their neighborhoods.

Berliner (2005) convincingly described poverty as a major factor that troubles classroom management in the U.S. According to authoritative statistics cited by

N.L. Gage, *A Conception of Teaching*,
DOI: 10.1007/978-0-387-09446-5_7, © Springer Science+Business Media, LLC 2009

Berliner, poverty is more widespread and longer-lasting in the U.S. than in comparably prosperous countries. And, in the U.S., poverty is associated with low scholastic achievement, especially among African-American and Hispanic students. As is shown in Tables 7.1 and 7.2, the average achievement of students from low-income African-American and Hispanic families is substantially lower than that of middle-class white American students and of students in other countries. Poverty-afflicted family environments and neighborhood environments prevent the realization of the genetic potential of students from low-income families, while middle-class students are able to achieve at their genetic potential (Turkheimer, Haley, Waldron, D'Onofrio, & Gottesman, 2003). The medical problems that impair scholastic achievement and opportunities beyond the school years occur more often and more severely among students from low-income families and neighborhoods.

Table 7.1 Mathematics scores (mean = 500) (*Source*: PISA* 2000: Lemke et al., 2003).

Country	Score
Japan	557
Korea, Republic of	547
New Zealand	537
Finland	536
Australia	533
Canada	533
United States Average Score of White Students	530
Switzerland	529
United Kingdom	529
Belgium	520
France	517
Austria	515
Denmark	514
Iceland	514
Sweden	510
Ireland	503
Norway	499
Czech Republic	498
United States Average Score	493
Germany	490
Hungary	488
Spain	476
Poland	470
Italy	457
Portugal	454
Greece	447
Luxembourg	446
United States Average Score for Hispanic Students	437
United States Average Score for African American Students	423
Mexico	387

*PIMA = Program on International Student Achievement *Source*: Berliner, 2005, p. 20

Table 7.2 Literacy scores (mean=500) from PISA* 2000 (Lemke et al., 2003)

Country	Score
Korea, Republic of	552
Japan	550
United States Average Score of White Students	538
Finland	538
United Kingdom	532
Canada	529
New Zealand	528
Australia	528
Austria	519
Ireland	513
Sweden	512
Czech Republic	511
France	500
Norway	500
United States Average Score	**499**
Hungary	496
Iceland	496
Belgium	496
Switzerland	496
Spain	491
Germany	487
Poland	483
Denmark	481
Italy	478
Greece	461
Portugal	459
United States Average Score for Hispanic Students	449
United States Average Score for African American Students	445
Luxembourg	443
Mexico	422

*PIMA=Program on International Student Achievement
Source: Berliner, 2005, p. 21

Accordingly, the test scores of students from low-income families and neighbor-hoods.in the U.S. are substantially lower than that of middle-class American students and middle-class students in other countries. Poverty-afflicted family and neighborhood environments prevent the realization of the genetic potential of students from low-income families, while middle-class students are able to achieve at their genetic potential, Also the medical problems that impair scholastic achievement and employment opportunities beyond the school years occur more often and more severely among students from low-income families and neighborhoods.

Most important, as Berliner states, seemingly modest *decreases* in the poverty level of low-income families result in improved student behavior and scholastic achievement. Poverty seriously restricts the potential of school reform for improving how students behave and how much they learn. He describes a study (Dearing, McCartney, & Taylor, 2001), that used, as a measure of poverty, the ratio of income-available to the [financial] needs faced by a family. A ratio of 1.00 means that the family is just making it, that its family income and needs such as housing, food, transportation, etc., are matched. A ratio of 3.00 would be more like that of a middle-class family, and a ratio of 0.8 would indicate poverty of some magnitude. A large and reasonably representative sample of poor and non-poor families were followed for 3 years, and their income-to-needs ratios were computed regularly, as were their children's scores on various social and academic measures. It was found was that, *as poor families went from poor to a lot less poor, for whatever reasons, their children's performance began to resemble that of the never-poor children with whom they were matched* (Berliner, 2005, p. 44).

Berliner's argument implies that one effective, and hitherto well-nigh unused, way to improve schools and achievement in the U.S. is to reduce the poverty of low-income families and their children. In the words of Dearing et al. (2001),

> Nearly 17% of children in the United States live in poverty (U.S. Bureau of the Census, 1999); thus, the importance of these findings, especially with respect to risk and prevention, is great from an incidence validity perspective (see Fabes, Martin, Harish, & Updegraff, 2000). The findings of the present study showed that naturally occurring decreases in family income-to-needs were associated with worse developmental outcomes for children from poor families. *Conversely, naturally occurring increases in family income-to-needs were associated with better developmental outcomes for children from poor families.* [italics added] (p. 1791)

Poverty and the Superintendency

Consistent with Berliner's portrayal of the way poverty affects classroom management is the description by Merrow (2004) of the difficulties of obtaining successful superintendents of big-city schools, such as those of Tucson, St. Louis, Houston, and Pittsburgh, where many of the students come from low-income families. As Merrow described it,

> Typically, a new superintendent arrives in a city, hailed as the answer to every problem – low test scores, poor attendance, embarrassing graduation rates. When change does not occur overnight, or perhaps at all, disappointment sets in. The superintendent departs for the next school district, and the cycle begins anew. (p. B-1)

This portrayal of a problem associated with classroom management points at the realities. Where poverty surrounds the parents and their children, classroom management has rough going.

Classroom Management in General

Fortunately, although poverty is widespread in the U.S., it is far from universal. Classroom management, along lines that research has found to be effective, has been able to provide classroom environments that foster learning and achievement. It is that kind of classroom management to which we now turn.

In its function as a prerequisite of teaching, classroom management seeks to optimize the amount of classroom time available for teaching's *processing* of *content* in ways suitable to *students' cognitive capabilities*. Conversely, classroom management seeks to minimize classroom time spent in ways that interfere with teaching and learning. In the minimizing function, classroom management seeks to prevent students from engaging in counterproductive activities, such as irrelevant conversation with other students, disruptive noise-making, wasted time, and distractive movement – activities that subtract from the time available for the teaching's processing the content to be learned and for the students' learning of that content. We turn now to each of these major aspects of classroom management.

Instructional Time

The study of classroom management has been formulated in terms of the concept of *instructional time*, or the ways in which time is used or misused in the classroom. A body of theoretical and empirical literature on the variables relevant to instructional time has emerged. The concept of *instructional time* provides an explanation for the positive correlation between the many kinds of classroom management and student achievement. Classroom management deals with techniques that minimize the amount of time spent in making transitions from one subject or activity to another and the amount of time spent in housekeeping and logistical activities, such as passing papers around or getting materials ready.

The instructional-time approach stems from influential publications by Carroll (1963, 1985, 1989), who proposed that three of five major factors in school learning – the student's *aptitude, opportunity to learn,* and *perseverance* – could be expressed in terms of time. Only the fourth factor, *ability to understand instruction,* and the fifth factor, *instructional quality,* could not be so expressed.

1. *Student aptitude.* The *amount of time* ordinarily required by a student for learning content in some domain, such as mathematics or science, other things being equal.
2. *Opportunity to learn.* The *amount of time* allowed for student learning by a program, school schedule, or teacher.
3. *Perseverance.* The *amount of time* a student is willing to spend on learning a task or unit of instruction, other things being equal.
4. *Ability to understand instruction.* In interaction with the method of instruction, especially in situations where the learner is left to infer for himself the concepts and relationships to be learned.
5. *Quality of instruction.* A variable that could not be expressed in terms of time but is considered to require that students be told clearly what they are to learn, that students must be put into adequate contact with learning materials, and that steps in learning must be carefully planned and ordered. (p. 714).

Carroll's model gave birth to a productive "instructional-time movement" in education, resulting in, among others, managerial applications to such variables as length of the school year and the school day. It was an important part of the basis for Mastery Learning (Bloom, 1968), an influential model of teaching during the 1970s. Other applications took the form of significant research on instructional time, that is, on ways in which teachers could use classroom management so as to optimize the use of classroom time and, thus, the level of student achievement (see, e.g., Ben-Peretz & Bromme, 1990; Berliner, 1987; Denham & Lieberman, 1980; Fisher & Berliner, 1985). But Gage (1978) pointed to a significant gap in Carroll's conception of time in the classroom:

> Academic learning time, in the form of allocated and engaged learning time is, in a sense, *a psychologically empty quantitative concept.* [italics added] We need better analyses of how that time is filled, of what learning processes go on during academic learning time. (p. 75)

Carroll (1989) replied:

> The problem is, of course, that although we can measure time – certainly elapsed time, and possibly "academic learning time" or "time on task," *we cannot meaningfully measure what goes on in the head of the student during that time* [italics added], or insure in any way that what goes on in the student's head is addressed to learning. (p. 27)

Studying Students' Thought Processes

Here Carroll was writing like a behaviorist. He was overlooking well-established methods for studying "what goes on in the head of the student" – methods developed within the cognitive psychology that became more prominent than behaviorism during the last 40 years of the twentieth century. E. D. Gagné, Yekovich, & Yekovich (1993) described five of the methods used in cognitive psychology for studying mental processes: (a) response latencies (how long it takes for a person to respond to a stimulus), (b) eye fixations (what part of a stimulus a person is looking at), (c) verbal reports (what a person says he is or was thinking at a certain time), (d) sorting (how a person sorts or classifies a set of objects), and (e) free recall (what a person can recall after being presented with certain stimuli).

By 1986, research on students' thought processes had flourished enough to warrant a chapter titled "Students' Thought Processes" (Wittrock, 1986b) in the third edition of the *Handbook of Research on Teaching* (Wittrock, 1986a). Wittrock began that chapter with a list of 12 kinds of students' thought processes:

> The recent research on students' thought processes studies the effects of teachers and instruction upon the student perceptions, expectations, attentional processes, motivations, attributions, memories, generations, understandings, beliefs, attitudes, learning strategies, and metacognitive processes that mediate achievement. (p. 297)

Particularly active in exploring students' thought processes were Marx and Winne (Marx & Winne, 1987; Winne, 1982, 1987, 1995; Winne & Marx, 1983, 1987). Answers to Carroll's methodological question – how to "study what goes on in students' heads" – took various forms. Especially studied by them was the question,

Do students actually seek cues for their learning during the teacher's discourse? On this question, Winne and Marx (1983) wrote

> Do students seek out signals about cognitions? Our answer showed that students notice cues for cognitions that their teachers use during instruction. However, this does not demonstrate that students actively look for cues. One way to show that students seek out cues is to locate places in lessons where teachers do not claim that a cue was used. Then, if students tell us that they observed a cue there nonetheless, we have evidence that they actively look for cues.
>
> Occasionally, we stopped the replay of a lesson for a teacher because we thought a cue had been given but had not been identified. On some of these occasions, the teachers said that they were not sure that they really used a cue at that point. On others, they were sure that a cue had been used but they gave ambiguous descriptions about the cue that they were trying to guide students to apply.... When we interviewed students, we replayed the videotape, stopping it at all these types of situations. The evidence supporting our contention that students are active seekers of cues comes from their responses to incidents where their teacher either was ambiguous about whether a cue was delivered or was certain that no cue had been used. (pp. 274–275)

The significance of the finding that students use cues in the teacher behavior they witness is that it points to the possibility of using students' thought processes as the basis for conjectures as to what is going on during instructional time. Such conjectures can be illustrated by holding that (a) students are paying attention to what the teacher is saying, (b) they understand the curriculum-relevant features of what she is saying and what they have read, (c) their participation in the "language game" of Bellack, Kliebard, Hyman, and Smith (1966) enables them to respond, either audibly or silently, to the teacher's solicitations, and (d) in formulating their responses they use cognitive processes of the kind reviewed by Wittrock (1986b).

Categories of Instructional Time

As the study of instructional time developed, it became clear that instructional time could be categorized. Berliner (1990) formulated the following:

> *Allocated time*: "The amount of time a state, district, school, or teacher provides the student for instruction" in a given subject. (p. 4)
> *Engaged time.* "[T]he time that students appear to be paying attention." (p. 4)
> *Time-on-task.*"[E]ngaged time on particular learning tasks." (p. 5)
> *Academic learning time.* "[T]hat part of allocated time in a subject-matter area.... in which a student is engaged successfully in the activities or with the materials to which he or she is exposed, and in which those activities and materials are related to educational outcomes. [i.e., are in instructional alignment] with assessments of outcome." (p. 5)
> *Transition time*: "[T]he noninstructional time before and after some instructional activity." (p. 5)
> *Waiting time.* "[T]he time a student must wait to receive some instructional help." (p. 6)
> *Pace.* "[T]he amount of content covered during some time period." (p. 6)
> *Off-task time.* To Berliner's list we add a kind of time that classroom management should minimize: time spent in such ways as the aforementioned irrelevant talk between students, disruptive noise-making, and distracting movement.

In terms of the definitions of these time variables, our conjectures would be that student achievement will be

– *positively correlated* with amounts of allocated time, engaged time, time-on-task, academic learning time, and pace, and

– *negatively correlated* with amounts of transition time, waiting time, and off-task time.

The implication is, of course, that the teacher should aim classroom management at increasing the *positively-correlated-with-achievement* kinds of time and at reducing the *negatively-correlated-with-achievement* kinds of time. Some research-based findings on classroom management can be translated into "teacher should" statements.

Classroom Management in Elementary Schools

The following examples of these findings in elementary school classrooms were developed by Brophy and Evertson (1976) and paraphrased by Gage (1978, p. 39):

Teachers should have a system of rules that allow students to attend to their personal and procedural needs *without* having to check with the teacher.

Teachers should move around the room a lot, monitoring students' seatwork *and* communicating to their students an awareness of their behavior, while also attending to their academic needs.

When students work independently, teachers should insure that the assignments are interesting and worthwhile yet still easy enough to be completed by each third grader without teacher direction.

Teachers should keep to a minimum such activities as giving directions and organizing the class for instruction. Teachers can do this by writing the daily schedule on the board, insuring that students know where to go, what to do, etc.

In selecting students to respond to questions, teachers should call on a child by name *before* asking the question as a means of insuring that all students are given an equal number of opportunities to answer questions.

With less academically oriented students, teachers should always aim at getting the child to give some kind of response to a question. Rephrasing, giving cues, or asking a new question can be useful techniques for bringing forth some answer from a previously silent student or one who says, "I don't know" or answers incorrectly.

During reading-group instruction, teachers should give a maximal amount of brief feedback and provide fast-paced activities of the "drill" type.

Classroom Management in Secondary Schools

Emmer and his co-workers (Emmer, Sanford, Clements, & Martin, 1982; Emmer, Evertson, Sanford, Clements, & Worsham, 1984) developed comparable findings concerning classroom management for junior and senior high school teachers

Examples of the results of their research are pieces of advice to teachers with such titles as "keys to room arrangement," "planning your classroom rules," and "monitoring student work in progress."

On the important problem of managing inappropriate behavior, Emmer et al. (1984) offered the following recommendations:

1. When the student is off task – that is, not working on an assignment – redirect his attention to the task: "Robert, you should be writing now." Or, "Becky, the assignment is to complete all the problems on the page." Check the students' progress shortly thereafter to make sure they are continuing to work.
2. Make eye contact with or move closer to the student. Use a signal, such as a finger to the lips or a head shake, to prompt the appropriate behavior. Monitor until the student complies.
3. If the student is not following a procedure correctly, simply reminding the student of the correct procedure may be effective. You can either state the correct procedure or ask the student if he or she remembers it.
4. Ask or tell the student to stop the inappropriate behavior. Then monitor the student until it stops and the student begins constructive activity. (p. 100)

Another realization about classroom management research is that, however obvious some of its findings may seem, many teachers at all grade levels have failed at classroom management. And they have been helped (e.g., Anderson, Evertson, & Brophy, 1979) to do a better job when given training based on the findings of research showing how more effective classroom managers behave differently from those who are less effective.

Avoiding Biases toward Students

Research has shown that teachers, usually unintentionally, exhibit biases of various kinds concerning students – biases that reduce students' ability to stay on task. Accordingly, avoiding such bias is also a significant aspect of classroom management.

The bias may stem from teachers' unwittingly differentiating disadvantageously against students on the basis of their gender, or ethnicity, or socio-economic status. Perhaps most difficult bias to avoid is that against low-achieving students.

Biases show up when students get treated differently by teachers in such ways as how often students are called upon,
how teachers react to students' responses,
how long teachers will wait for students to respond to a question,
how often students in different parts of the classroom – say, the rear and the sides – may be called upon,
how often teachers give justified praise or reproof to students.

Since the bias is usually unintentional, teachers tend to be unaware of it. If so, merely mentioning bias may tend to reduce it. Another possibly helpful practice calls upon Teacher A to invite Teacher B to observe in Teacher A's classroom, and perhaps count the occurrences of any of the above-mentioned biases. Analyzing such data might make Teacher A aware of her biases.

The concept of *instructional time* provides an explanation for the positive correlation between the many kinds of classroom management and student achievement. Classroom management deals with techniques that minimize the amount of time spent in making transitions from one subject or activity to another and the amount of time spent in housekeeping and logistical activities, such as passing papers around or getting materials ready.

Chapter 8
Integrating the Conceptions

The notion of a single theory that does full justice to the phenomenon being explained is, of course, attractive but infeasible for teaching. An example of a single adequate theory is Isaac Newton's second law of motion; it embraces only three variables (force, mass, and acceleration), related as $f = ma$, or force equals mass times acceleration. Even so, it held only for frictionless surfaces and motion in a vacuum. In the everyday world, without the vacuum and with friction, matters get more complicated.

Typically, scientific theories refer to relationships between many variables. In the behavioral sciences, theories of complex phenomena are abundant. Behavior theory (Skinner, 1953) relates drive, behavior, reinforcement, extinction, response generalization, and operant conditioning. Cognitive learning theory (Levine, 1975) deals with short-term memory (working memory), long-term memory, and its rehearsal, decoding, and retrieval processes. Cognitive-abilities theory (Carroll, 1993; Detterman, 1994; Gardner, 1983; Neisser, 1998) deals with general ability (g), group abilities, special abilities, and IQ change. Cognitive dissonance theory (Festinger, 1957; Harmon-Jones & Mills, 1999) speaks of relationships between dissonance ratio, psychological comfort, aversive consequences, self-affirmation, and dissonance magnitude.

Sub-Theories

Teaching is, of course, immeasurably more complex than motion in a frictionless vacuum. Because the process of CDR teaching comprises four components (structuring, soliciting, responding, and reacting), we begin by formulating separate sub-theories, for each of those components. The subsequent integration of the sub-theories of process, the sub-theories of content, the sub-theories of cognitive capability and motivation, and the sub-theories of classroom management will constitute a theory of teaching.

The philosopher of science Hempel (1965) expressed this multi-explanation conception of theory in the following terms:

N.L. Gage, *A Conception of Teaching*,   123
DOI: 10.1007/978-0-387-09446-5_8, © Springer Science+Business Media, LLC 2009

[A] deductive-nomological [D-N] explanation is not conceived as invoking only one covering law; and our illustrations show how indeed many different laws may be invoked in explaining one phenomenon. A purely logical point should be noted here, however. If an explanation is of the form (D-N), then the laws L_1, L_2, ..., L_r invoked in its explanans [explaining ideas] logically imply a law L which by itself would suffice to explain the explanandum [thing to be explained] event by reference to the particular conditions noted in the sentences C_1, C_2, ..., C_k. This law L is to the effect that whenever conditions of the kind described in the sentences C_1, C_2, ..., C_k are realized then an event of the kind described by an explanandum-sentence occurs. (p. 346)

An Illustrative Theory Consisting of Sub-Theories[*]

A multi-component process can illustrate the conception of separate sub-theories – one for each of the components of teaching. This is the multi-component theory that explains a single phenomenon: how an automobile is made to move as the result of a sequence of four component processes:

1. *Delivery of Fuel and Air*. A gasoline-powered automobile engine has a fuel-injection system that sprays a precise mixture of gasoline and air into the space between each piston and the cylinder's walls. The covering law is derived from *hydrodynamics theory* that explains how fluid flow (vaporized gasoline and air) results in a combustible mixture of gasoline and air in the cylinder.
2. *Ignition*. A precisely timed electrical spark ignites the vaporized gasoline and air. The covering law is derived from *combustion theory* that would explain how ignition produces the combustion of the mixture that creates heat, causing the gaseous mixture to expand and exert pressure on a piston. Each other piston then moves because of its connection to one of the rotating cranks on the crankshaft.
3. *Reciprocation*. The pistons' up-and-down or back-and-forth motion is transformed by the connecting rods into the circular motion, or rotation, of the crankshaft. The covering law is derived from the *theory of mechanics* that would explain the motion of levers connected to a shaft.
4. *Transmission*. The circular motion of the crankshaft is transmitted by the differential – a system of gears and shafts – to the axles of the wheels, causing the axles and wheels to rotate and the car to move.

The covering law here is derived from *theory of mechanics* that would explain how a rotating rod can have its mechanical energy converted into the rotation of an axle perpendicular to the rotating wheel.

Now we go on to develop a sub-theory of the *process* of teaching. Next, we do the same for the *content* of teaching. Then sub-theories of *cognitive capabilities* and *classroom management* are developed. Finally, the eighth chapter integrates the sub-theories of process, content, cognitive capabilities, and classroom management to formulate a theory of teaching.

---

[*] Written in collaboration with Thomas Burrows Gage.

Scheme of Presentation of Sub-Theories

The following presentations of sub-theories first provide an ***example of the phenomenon to be explained*** and ***then a covering law*** explaining that phenomenon. The term *example* signifies that the pedagogical phenomenon has not necessarily been as well established as if, for instance, it had been demonstrated by successive valid meta-analyses. That requirement could not have been met, in view of the lack of sufficient research-based evidence. Instead, only an example of the phenomenon is presented, and the proposed covering law is offered to indicate what the explanation could have been if the phenomenon had been well established.

The term *covering law* was used by Hempel (1965, pp. 345–346) to refer to the deductive-nomological explanation of the phenomenon. Unlike his explanations of phenomena in the physical sciences, the covering laws used in the present theory of teaching are not strong deductive-nomological laws. Rather, they may be a logical law, such as a valid syllogism. Or they may be a universally accepted commonplace, such as the desirability of rationality. Or they may be well-established empirical generalizations, such as the positive correlation between students' socioeconomic status and their academic achievement. Each of these possibilities may be used to explain a phenomenon that occurs in teaching. The explanation of the phenomenon is deduced from the covering law.

Sub-Theories of the Process of Teaching

Sub-Theories of Structuring

We turn now to sub-theories for each of the four components – structuring, soliciting, responding, and reacting – of the process of CDR teaching. Each of these can be the subject of a sub-theory. We begin with Bellack's definition.

> *1.1 Structuring (STR). Structuring moves serve the function of setting the context for subsequent behavior by (1) launching or halting-excluding interactions between teacher and students, and (2) indicating the nature of the interaction in terms of the dimensions of time, agent, activity, topic and cognitive process, regulations, reasons, and instructional aids. A structuring move may set the contest for the entire classroom game or a part of the game.* (Bellack, Kliebard, Hyman, & Smith, 1966, pp. 16–17)

A concept similar to structuring was defined by Smith, Meux, Coombs, Nuthall, and Precians (1967):

> *The Beginning of a Venture.* An utterance or part of an utterance containing an explicit indication (announcement or proposal), usually by the teacher, that a particular topic is to be considered [compare Bellack et al.'s Structuring]. Such an announcement is usually followed by a question which initiates discussion of the proposed topic or by an invitation to speak [compare Bellack et al.'s Soliciting] on the topic. (p. 13)

Table 8.1 Examples of brief structuring

Process	Content
1. "Let's turn now to ¼	*the difference between nouns and verbs."*
2. "We mustn't overlook ¼	*how dividing by fractions differs from dividing by whole numbers."*
3. "But the picture changes when we turn from ¼	*evaporation to condensation."*
4. "I want you to appreciate ¼	*the horrors of what slavery meant to those who now are called African-Americans."*

Examples of brief structuring. Everyone who has been a student has experienced a teacher's structuring. We readily think of examples, embracing both process and content, of what teachers have done in "setting the context for subsequent behavior or performance." Table 8.1 presents brief examples.

The length and duration of teacher structuring can be much larger than these examples as teachers "set the context," "convey an implicit directive," "launch classroom discussion in specified directions," and "focus on topics, subjects or problems to be discussed, or procedures to be followed."

Functions of Structuring

Whether structuring is as brief as these examples or much longer, it has several functions in CDR teaching: focusing attention, modifying content, agenda setting, and guiding students' cognitive processes.

Focusing Attention. Helping students understand the reasoning underlying the immediate part of the curriculum – and its corresponding educational objectives – is an important function of structuring. Accordingly, at an early point in the consideration of that subject matter, the teacher makes statements about the importance to her students – and the reasons for that importance – of the subject matter to be studied. At the beginning of each new topic, its importance should be explicated, or left implicit, to the extent judged appropriate by the teacher. That importance should stem from the teacher's perception of the needs, usually intellectual needs, of the students and the needs – economic, political, or social – of the society in which they live. As each subsequent topic in the content comes to the fore, the teacher should judge whether the importance of the topic should be pointed out or whether its importance is obvious enough to make importance-explicating unnecessary.

How well the teacher succeeds in focusing attention shows up in what Carroll (1989) called *student perseverance*, or "the amount of time a student is willing to spend on learning a task or unit of instruction, other things being equal" (p. 26). More able students need to exhibit less perseverance in studying a given topic because they learn more rapidly and need less time to achieve the objectives of the teaching. The focusing of attention is manifested in the "system of human memory" set forth by Atkinson and Shiffrin (1968):

Our ... categorization divides memory into three structural components: the sensory register, the short-term store, and the long-term store. Incoming sensory information first enters the sensory register, where it resides for a very brief period of time then decays and is lost. The short-term store is the subject's working memory; it receives selected inputs from the sensory register and also from the long-term store. Information in the short-term store decays completely and is lost within a period of about 30 seconds, but a control process called rehearsal can maintain a limited amount of information in this store as long as the subject desires. The long-term store is a fairly permanent repository for information, information which is transferred from the short-term store. (pp. 90–91)

Example 1 of the phenomenon to be explained: Newby (1991) found that, among 30 first-year elementary school teachers, "those classrooms in which there was a higher incidence of giving reasons for the importance of the task ¼ showed a higher rate of on-task behavior" (p. 199).

The covering law: As part of the attention-focusing process, giving justifications for expending effort is effective because of the students' rationality.

Modifying the Content: Taking into account the students' personalities (e.g., prior knowledge, interests), the teacher can modify the content so as to optimize its physical properties (e.g., size, color, motion), aesthetic properties (e.g., attractiveness, symmetry), psychological properties (e.g., relevance to personal, social, and emotional needs), and intellectual properties (e.g., difficulty, complexity, logicality).

Example 2 of the phenomenon to be explained: *Sesame Street*, a television show for children (Fisch & Truglio, 2001), improved children's readiness for school and their subsequent school achievement. Its material was explicated in detail, appealing enough to attract children away from competing activities, concrete and explicit, child-centered and child-relevant, repetitive and reinforcing, providing models and identification, and inviting participation. The visual and sound materials exhibited color, movement, strangeness joined with familiarity, an appropriate intellectual level, and humorous juxtapositions.

The covering law: Modifying content is aimed at improving the fit between a student's personality (e.g., abilities, prior knowledge, interests, values) and the characteristics of the material to be learned (e.g., difficulty and relevance to students' interests).

Because of differences in their personalities, e.g., their abilities and interests, students are likely to differ in their motivation to learn any specific content. The teacher's task becomes one of seeking, through structuring, to describe the material to be learned so as to fit the personalities of as many students as possible. Hence, modifying the content, or appropriate description of the material, can focus the attention of students at any level of capability.

Agenda Setting. The teacher determines what content is to be learned and the successive steps of the process by which it is to be learned.

Example of the phenomenon to be explained: Observation in classrooms indicates that students almost always accept the teacher's agenda setting, namely, defining the content to be learned and the classroom processes through which that learning is to be brought about. When the teacher signals ("Let's turn now to...") that her class should focus on a certain aspect of the content – particular facts, concepts, principles, skills, problems, and more – ("by engaging in") – some particular process, the students typically accept the teacher's authority to make such decisions.

The covering law: Role theory (e.g., Biddle, 1979; Johnson & Johnson, 1995; Newcomb, 1950) explains this student acceptance as resulting from the superior status of the teacher – a status derived from the teacher's social position in the classroom, the school, the community, and the society – and the role (expected behavior) that accompanies that position. Students' perceptions of the teacher's role result in students' expectations that the teacher will take responsibility for guiding students as they seek to achieve objectives specified in the curriculum.

Example 1 of the phenomenon to be explained: The teacher can guide students' cognitive processes by suggesting that students focus on some content – an event, phenomenon, problem-solution, or other intellectual entity relevant to the objectives of the lesson. The teacher can suggest that students monitor their own intellectual processes, that is, use *metacognition*, by asking themselves such questions as "Do I really understand what the teacher or the book has said?" or "Can I apply this principle?" or "Should I review this material again?"

Example 2 of the phenomenon to be explained: Metacognitive training can improve mathematical reasoning, as illustrated by an experiment (Kramarski & Mevarech, 2003a, b) in which eighth-graders received metacognitive training consisting of three sets of self-addressed metacognitive questions: comprehension questions, strategic questions, and connection questions. As Kramarski and Mevarech (2003b) put it,

> The comprehension questions were designed to prompt students to reflect on a problem before solving it.... The strategic questions were designed to prompt students to consider which strategies were appropriate for solving or completing a given problem or task and for what reasons .... The connection questions were designed to prompt students to focus on similarities and differences between the immediate problem or task and problems or tasks that they had already completed successfully. (p. 286)

Compared with an untrained control group, the trained group showed improved achievement of mathematical reasoning.

Example 3 of the phenomenon to be explained: Another version of metacognitive effects occurs when students are advised to use "chunking" to facilitate their ability to remember a series of numbers, letters, or words. The chunking breaks up a series of, say, 15 items (words, numbers, names, etc.) into, say three groups of five items, i.e., small groups that are easier to remember.

The covering law: Metacognition, or students' monitoring of their own cognitive processes, improves achievement by providing the student with covert practice, self-guidance, and rewards.

Structuring as Lecturing

Structuring lends itself to further analysis through the kinds of research that have been done on lecturing. The teacher, in structuring the discourse that sets the stage for the subsequent components of the teaching cycle, may give a brief lecture – typically, an uninterrupted monolog.

The Duration of Structuring. How long structuring should go on depends on the maturity of the students and the characteristics of the content. Most writers on lecturing deal with lectures aimed at relatively mature audiences, such as undergraduates and graduate students. In the present context, we consider the teacher's structuring to be carried out as a shorter lecture at the elementary and secondary school levels.

Although both structuring and lecturing are teacher monologs, they should be differentiated. A lecture typically lasts longer – for, say, up to a whole 50-min class period – and presents information for students to *remember* and *comprehend*, along lines described in the *Taxonomy for Learning, Teaching, and Assessing: A Taxonomy of Educational Objectives* (Anderson, Krathwohl, Airasian, Cruikshank, Mayer et al., 2001).

Structuring moves tend to be shorter than lectures. Structuring moves occur in a smaller proportion of the discourse than any of the other three types of moves (soliciting, responding, and reacting). or less than 5% of all moves. But, as Bellack et al. (1966) noted, when

> these data are considered *in terms of lines spoken, structuring accounts for a much greater percentage of the teacher's discourse* [italics added]. This is to be accounted for by the fact that structuring moves are generally longer than other types of moves. (The mean length of structuring is 9.1 lines of transcript, in contrast to 3.5 lines for reacting, 2.0 lines for responding, and 1.9 lines for soliciting). (p. 153)

Example of the phenomenon to be explained: Everyday experience with children and adolescents of varying ages indicates that younger children can typically pay attention to oral discourse, other things (such as the interest and importance of the content) being equal, for shorter periods than older children and adolescents. Older individuals, at least up to the ages of about 60, have longer attention-duration abilities. But uncommonly interesting phenomena, such as the children's television program *Sesame Street*, show that attention duration is determined by the characteristics of the content (such as the aforementioned "color, movement, strangeness joined with familiarity, intellectual level, and humorous juxtapositions,") as well as the maturity of the audience.

The covering law: Attention-duration is determined by the maturity of the students and the characteristics of the content.

So our conception of structuring as a part of the teaching cycle must take into account the limits on structuring duration imposed by the level of maturity of elementary- and secondary-school students. Taking into account the typical attention-duration of students of different ages, the teacher adjusts the duration of structuring – the amount of time during which the teacher structures, with only slight or no interruption. So, holding content characteristics constant, the structuring typically must be relatively short (say, 0.25–2.0 min) in the lower school grades, and this attention-duration increases gradually as grade level increases to, say, 20 min in grade 12. Only at the college and graduate-school levels does the traditional 50-min lecture hold student attention.

The Comprehensibility of Structuring. The most important feature of structuring is its comprehensibility. Often called "clarity," it also betokens intelligibility, or the degree to which the structuring achieves its purpose of conveying to students its message about the nature of what the next episodes of the teaching will be concerned with.

Comprehensibility is measured by the degree to which students to whom the structuring is addressed exhibit, first, appropriate cognitive processes during the structuring and, second, the success of students in answering relevant oral or printed questions. Much research on the importance of the features of structuring has examined the opinions of college students concerning what features of lectures are important to them. Summarizing such research, Hativa (2000) found that the feature of lectures that showed up most frequently and strongly in studies of college students' opinions was *clarity*. It seems safe to assume that, in view of clarity's meaning, students at lower grade levels would agree.

Communicative Logic and Comprehensibility. Structuring's *logicality,* or conformity to the rules of logic, is an important factor in comprehensibility. Everyday experience tells us that comprehensibility often depends on the logic of what is said. Everyone has experienced hearing discourse that was incomprehensible because it was illogical. Ausubel (1963) wrote as follows on the concept of logical meaning:

> It implies that the learning material per se consists of possible and nonarbitrary relationships that *could* be nonarbitrarily incorporated on a nonverbatim basis by a hypothetical human cognitive structure that had the necessary ideational background and degree of readiness. This criterion of logical meaning applies primarily to the attributes of the material itself. If it (the material) possesses the characteristics of nonarbitrariness, lucidity and plausibility, then it is, by definition, also relatable to the aforementioned hypothetical structure. (p. 39)

Examples of the phenomenon to be explained: Studies by Needels (1984, 1988) and Huh (1985) investigated whether structuring becomes less comprehensible as its illogicality increases. Needels studied the teacher as the exhibitor of logical discourse. Her definition of discourse resembled that of Bellack et al. (1966): "lecturing, asking questions of students, reacting to students' responses, and responding to students' questions and comments" (Needels, 1988, p. 509).

She applied the general conception of logical validity in a study of relationships between logic in teaching and the achievement of students in comprehending and appreciating what was taught. The logic entailed here was *not* formal logic of the kind found in philosophy, science, and mathematics, and applied to teaching by Ennis (1969). Rather, she studied "communicative logic" as formulated by Grice (1975):

> It is a commonplace of philosophical logic that there are, or appear to be, divergences in meaning between, on the one hand, at least some of what I shall call FORMAL devices— $\sim, , , > , (x), E(x), fx$ (when these are given a standard two-value interpretation)—and, on the other, what are taken to be their analogs or counterparts in natural language—such expressions as *not, and, or, if, all, some (*or *at least one), the.* Some logicians may at some time have wanted to claim that there are in fact no such divergences; but such claims, if made at all, have been somewhat rashly made, and those suspected of making them have been subjected to some pretty rough handling. (p. 41)

So there must be a place for an unsimplified, and so more or less unsystematic, logic of the natural counterparts of these devices; this logic may be aided and guided by the simplified logic of the formal devices but cannot be supplanted by it; indeed, not only do the two logics differ, but sometimes they come into conflict; rules that hold for a formal device may not hold for its natural counterpart (p. 43).

Grice's argument makes room for the analysis of natural language in terms of its "communicative logic." The natural language studied by Needels (1988) was that of teachers in structuring for students what subsequent discourse will be about. In considering such discourse, Needels used Grice's "Cooperative Principle": "Make your conversational contribution, such as is required, at the stage at which it occurs, by the accepted purpose or direction of the talk exchange in which you are engaged" (Grice, 1975, p. 45).

Grice also stated four "maxims" of the Cooperative Principle:

(a) Quantity

1. Make your contribution as informative as is required for the current purposes of the exchange.
2. Do not make your contribution more informative than is required.

(b) Quality

1. Do not say that which you believe to be false.
2. Do not say that for which you lack sufficient evidence.

(c) Relation, that is,

1. Be relevant.

(d) Manner, that is,

1. Avoid obscurity of expression,
2. Avoid ambiguity,
3. Be brief (avoid unnecessary prolixity),
4. Be orderly (Grice, 1975, pp. 45–46).

Needels (1988) used these four maxims of Grice's conception of communicative logic in a study of relationships between (a) logical variables in teaching and (b) the degree of students' comprehension of, and interest in, what was taught. She began by noting that classroom discourse differs from informal conversation in being more constrained as to the topic and the participant; in each case the teacher's choices dominate. Then she identified six Grice-implied variables, each expressed *negatively*, that is, in *illogical* form:

Manner:	1. Confusing syntax
Quantity:	2. Omission of necessary definition
	3. Omission of causal factors
Relation:	4. Irrelevance
Quality:	5. Incorrect use of words
	6. Incorrect causal relationship. (pp. 507–508)

The study recorded and transcribed what each of ten sixth-grade teachers said to their students during a 1-day lesson on light. The transcripts were analyzed twice: first, to obtain scores for the teacher's perpetration of each of the six kinds of *illogic*; and second, to determine whether the teacher subsequently, during the lesson, *corrected* each of the instances of illogic.

The students' scores on *pretests* of (a) scholastic aptitude and (b) prior knowledge of light were obtained. And *posttest* scores were obtained for student achievement of the lesson's objectives, namely, (a) comprehension of the lesson content and (b) interest in the content of the lesson.

The unit of analysis of the data was the test item. The results for the first analysis of teachers' initial, that is, *not self-corrected,* violations of the logic of classroom discourse showed "no consistency [across test items] in the direction (positive or negative)" of the correlations between the teachers' *illogicality* measures on the subject matter of a test item and their students' percentage of correct responses to that item.

But then the teachers were scored for their *subsequently (during the same presentation) corrected-by-themselves versions of their communicative illogic.* After such corrections, consistent across-items negative relationships (determined with a test of the significance of combined results) were found between (a) the teacher's self-corrected degree of illogicality on Confusing Syntax relevant to a given item of the lesson-comprehension test and (b) their *high-aptitude* students' percentage of correct responses on that item of the test. Apparently, only the high-aptitude students became confused by their teacher's *illogic* and understood that item of the test less well.

Also, the correlation between (a) frequency of illogical discourse and (b) the students' mean favorability of attitude toward the lesson content was determined. It suggested that "a greater degree of communicative illogic on the teacher's part was associated with less favorable student attitudes toward the lesson content" (Needels, 1988, p. 522).

A Second Study of Communicative Logic. Huh (1985) carried out a study similar to that of Needels. The main difference was that it was conducted in Seoul, South Korea, not the U.S. Huh also defined, in form, ten kinds of *illogic*: (a) confusing syntax, (b) vagueness, (c) verbal mazes, (d) incomplete explanation, (e) irrelevance, (f) gaps in definition, (g) incoherence, (h) inconsistency, (i) incorrect explanation, and (j) analytical errors.

Ten sixth-grade teachers taught their students two 1-hr lessons on seven topics about light: reflection, refraction, properties of light, lenses, absorption, prisms, and color. The audio-recorded lessons, transcribed and analyzed, yielded scores for each teacher on each of the ten kinds of illogic. The teachers' scores on the measures of *illogicality* differed significantly from one another and correlated negatively ($r = -.24$) with their students' total scores on two tests of lesson comprehension.

In short, these two studies suggest that structuring, like all oral communication, is comprehensible to the degree that it conforms to the principles of "communicative" logic.

The covering law: Valid communicative logic meets the listener's needs and expectations acquired through experience in everyday discourse.

The Communicability to Teachers of Comprehensibility-Affecting Actions

Telling teachers what actions, or behavioral variables, affect comprehensibility involves the distinction between high-inference and low-inference variables (e.g., Gage, 1969, p. 1452). The inference-level of a variable is the amount of inferring, i.e., deriving by reasoning or judging from premises or evidence – that the observer or auditor of teaching behavior must use in judging the occurrence or amount of that behavior variable. Some behaviors require relatively little inference; examples of these are the numbers of times a teacher calls on girls or boys and the frequency with which the teacher writes on a chalkboard. Other behaviors require a considerable inference from what is seen or heard in the classroom, such as the degree to which the teacher is partial versus fair, autocratic versus democratic, aloof versus warm, or the degree to which students are apathetic versus alert, or obstructive versus cooperative.

Thus inference-level becomes important in teacher education when oral or printed advice is given to teachers on how they should perform in order to act appropriately on the need for, say, comprehensibility. It becomes important in studying teaching when observers must judge, say, comprehensibility. A high level of inference causes the words presented to be susceptible to many interpretations, i.e., vague and ambiguous. A low level occurs when the words are explicit, readily interpreted, and precise as to their references to behavior.

Typically, high-inference features of teaching are evaluated with a rating scale, that is, "an instrument used to assign scores to persons or objects on some numerical dimension" (VandenBos, 2007, p. 769). Low-inference features are typically estimated by counting observers' tallies of the occurrences of the behavior being measured. It is sometimes held that high-inference instructions are the more effective for communicating "molar," or comprehensive, abstract aspects of action or behavior and that low-inference instructions are more effective for communicating "molecular," or specific, components of action or behavior.

High-Inference Variables. McConnell and Bowers (1979) furnished examples of high-inference variables; they included ratings of teacher clarity, teacher variability, teacher enthusiasm, teacher use of student ideas, teacher provision of opportunity to learn, and teacher task-orientation. Such ratings call for much inference from the terms used to the action designated, often resulting in ambiguity and confusion for the recipient of the advice.

Low-Inference Variables. When the amount of inference from the advice is low, determiners of comprehensibility are more unambiguously expressed, understood, and acted upon. The following list, drawn from McConnell and Bowers (1979), illustrates low-inference variables:

> Teacher statements were classified as affective, substantive, or procedural. Interchanges, particularly those relating to content, were classified as to entry and exit. An entry to an interchange could be a pupil question, a teacher question getting a student to elaborate, a teacher question intending to elicit pupil thought and divergent response, or a teacher question aimed at a specific response. The exits were based on teacher responses to the student portion of the interchange: acceptance, disapproval, or neutral. (p. 6)

Inference Level and Outcome Prediction. The inference level of teacher-action variables may be significant not only for the communicability of advice to, or observation of, teachers. It may also affect the value of a teacher-behavior variable for correlating with, or predicting, student achievement and attitude.

Example of the phenomenon to be explained: McConnell and Bowers (1979) compared high- and low-inference measures of teaching behaviors as to their value for predicting student achievement and attitude in 43 algebra classes. High-inference measures included ratings of the teacher's "enthusiasm" and "clarity," and low-inference measures included frequency counts of such behaviors as "problem-structuring" and "rebuking." Of the 87 correlations of the low-inference types of process measures with student achievement, 18% were statistically significant (i.e., probably not due to chance) while only 11% of the 72 high-inference correlations were statistically significant.

The covering law: The correlation of teaching variables with student-achievement measures is in part a function of the reliability of the measures of the teaching variables, and that reliability is a function of the level of inference required in judging the occurrence or amount of the teaching variables.

Sub-Theories of Soliciting

1.2. Soliciting (SOL). Moves in this category typically take the form of questions intended to elicit (a) an active verbal response on the part of the persons addressed; (b) a cognitive response, e.g., encouraging persons addressed to attend to something; or (c) a physical response. (Bellack et al., 1966, p. 18)

As the CDR model specifies, the next step in an episode of teaching is *soliciting.* The term "soliciting" is used, rather than its more common near-synonym *questioning,* because soliciting need not take grammatically interrogative form. "Although these [soliciting] moves may take all grammatical forms – declarative, interrogative, and imperative – the interrogative occurs most frequently" (Bellack et al., 1966, p. 18). A soliciting move may be a declarative sentence that begins with an emphasis (*Magellan* discovered America?) or ends with an upward lilt (17 + 13 = *30?*). Such forms of soliciting serve the same purposes as questions.

Goodlad (1984), in his observational study of 1,017 classrooms, found that "Less than 1% of [soliciting] time is devoted to open questions which call for more complex cognitive or affective responses [by students]" (Sirotnik, 1983, p. 20). Observations of CDR teaching indicate that teachers uniformly, in all grade levels and subject matters, ask about 90% of the questions in classroom discourse. Students do a correspondingly high percentage of the responding (Dillon, 1988).

Pedagogical moves occur in classroom discourse in certain cyclical patterns and combinations, which are designated teaching cycles. A teaching cycle begins either with a structuring move or with a solicitation that is not preceded by a structuring move. (Bellack et al., 1966, p. 19)

The question serves (a) to *focus* the students' attention on some part of the structuring, and (b) to *engage* the students in discovering how well they have understood the structuring.

Teacher Questions During Discussions. In discussions, students' utterances are longer than those in recitations, are aimed at fellow participants (classmates), and are mostly steered by the students themselves. Teachers often ask questions, or solicit, when they want students to participate in a discussion of a topic, issue, or phenomenon. Teachers' questions should obviously call forth students' responses.

But does a discussion occur? A discussion is usually intended to enable students, through give-and-take *among themselves,* to engage in problem-solving, sharing opinions, practicing self-expression, thinking for themselves.

Example of the phenomenon to be explained: In classroom discussions, as against recitations, the teacher's asking a question tends to *thwart* rather than foster the discussion, as is illustrated in Fig. 8.1 (Dillon, 1985). A teacher's questions during a discussion tend to make students' responses shorter and addressed toward the teacher rather than other students.

On the other hand, a teacher's "non-question" fosters the discussion, i.e., results in student utterances that last longer and are directed to the classmates, fellow participants in the discussion, and not solely to the teacher.

Dillon's finding is counterintuitive in that teachers often ask questions with the intent of fostering, rather than thwarting, discussion. They presumably reason that their questions will influence students to respond and thus to participate in the discussion. When that does not happen, teachers are discouraged from trying for a discussion again. Dillon's finding may explain why many teachers, after a few unsuccessful attempts, abandon the discussion method.

Covering law: The teacher's asking a question during a discussion immediately reinstates for the students the traditional teacher's role in a *recitation*, in which she typically asks questions to be followed by brief student responses addressed to the teacher.

On the other hand, a teacher's *non-question* utterance during a discussion is perceived by the students as making the teacher one-among-equals, that is, as having the same status as the students participating in the discussion, so that the students' utterances can be longer and directed toward other students. Non-question utterances of teachers during a student discussion can take the forms shown in Table 8.2.

Types of Questions. The teacher's question tells the students which of several kinds of learning they should have acquired as a result of the structuring. Blosser et al. (1973) classified questions according to whether they were *instructional* or *managerial.* We shall be concerned with only the former of these.

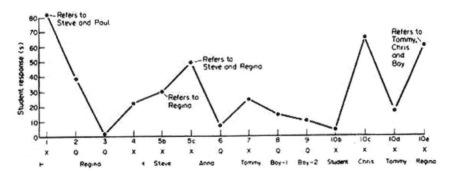

Fig. 8.1 After teacher questions (Qs), students' utterances become briefer, but after teacher non-questions (Xs), students' utterances are longer (Adapted from Dillon, 1985, p. 116).

Table 8.2 Types of non-questions usable by teachers seeking to foster discussion

1. Provide that participants formulate the question that now appears at issue in the discussion.
2. Utter a brief phrase, quietly exclaim feeling in reaction to what the speaker has just finished saying.
3. Emit some word or sound, indicate attentive interest in what the speaker has said or is in the process of saying.
4. By gesture or statement, pass the next turn at talk to another speaker.
5. Say nothing at all but maintain a deliberate, appreciative silence for 3–5 seconds or so.

Source: Dillon (1990, pp. 196–199, passim.)

As noted in Chap. 4, instructional questions can be classified according to the types of learning objectives formulated in the *Taxonomy for Learning, Teaching, and Assessing* (Anderson et al., 2001). That *Taxonomy* identifies two dimensions of educational objectives:

(a) *the Knowledge Dimension* (Factual Knowledge, Conceptual Knowledge, Procedural Knowledge, and Metacognitive Knowledge), and
(b) *the Cognitive-Process Dimension* (Remember, Understand, Apply, Analyze, Evaluate, and Create).

Questions asked by the teacher can be classified on the same two dimensions: (a) *the kind of knowledge* asked for and (b) *the kind of cognitive process* asked for. A cognitive objective of educational achievement consists of the pairing of a kind of Cognitive Process (remember, understand, apply, analyze, evaluate, and create) *with* a kind of Knowledge (facts, concepts, procedures, and metacognitions, that is, awareness of their own understandings of the subject matter). Thus, an educational objective might consist of *remembering* (cognitive process) the *multiplication table* (type of knowledge); or an *ability to evaluate*. And the teacher's soliciting, or questioning, takes one of these forms.

Relationships of Question-Level to Outcomes. One sub-theory of soliciting consists of explanations of relationships between (a) the type of question asked and (b) the type of learning affected. Such relationships should – according to the assumption that people learn what they have practiced – be positive. That is, students asked one kind of Process D Content question should be expected to achieve more of the intended Process D Content objective of the teaching than students not asked such questions.

The issue becomes salient in view of the often-reported impression that teachers' questions tend to be nearly always aimed at students' *remembering* of *facts* about what the students have read, heard, observed, or discussed. This means that teachers seldom ask questions aimed at one of the other five cognitive-process objectives of the Taxonomy (abilities to understand, apply, analyze, evaluate, or create). The discrepancy may mean that students are not stimulated enough to acquire these "higher" cognitive abilities, or ways of processing information in ways that are more complex than remembering. The conjecture underlying these concerns is that teachers who ask more of the higher cognitive-process questions will help students achieve the higher cognitive objectives to a greater degree.

A descriptive study by Giaconia (1987) led to her impression about how the kinds of questions asked affect the kinds of responses students make:

> Many of the findings reported for the relationship of student characteristics, question characteristics, and wait-time [how long the teacher waits after asking a question before calling on a student] characteristics with these response characteristics seemed to "make sense," and painted a picture of the *recitation* [italics added] as an activity that is reasonable and generally effective. That is, the nature of students' responses was highly related to the nature of teachers' questions. (p. 253)

Experiments Comparing Higher- and Lower-Order Questions. Winne (1979) reported on a careful review of 18 experiments and quasi-experiments intended to test the validity of that conjecture. (An *experiment* is an investigation of the difference of results from two or more *manipulated* treatments, such as the teacher's asking much more or much fewer higher-order questions, administered to two *randomly equivalent* groups.)

Of these 18 experiments, nine were "training experiments" in which the teachers, after being trained, "were free to use the skill(s) on which they were trained at their discretion in teaching" (p. 15). The other nine were "skills experiments" in which "the teachers' frequency and use of a skill is prescribed by the experimenter" (p. 15). Quasi-experiments are research in which the investigator cannot assign subjects at random to experimental or control groups and control or manipulate the independent variable but can determine how the dependent variable is measured" (VandenBos, 2007, p. 763).

Winne's review antedated the era of meta-analysis; that is, he used careful description and analysis of each of the experiments individually as the basis for his subsequent interpretation of the results. He also used now-obsolete (Kline, 2004) estimates of the statistical significance (non-chanceness) of each experiment's results, rather than the present-day, more widely accepted-as-appropriate, estimates of *effect size*, or the difference between the means of the experimental and control groups divided by the standard deviation of the control group.

Winne found that only 3 of the 18 experiments yielded statistically significant (that is, non-chance) results showing higher achievement for the classes receiving the higher-order questioning treatment, 11 yielded non-significant (that is, possibly due to chance) results, and 4 yielded uninterpretable results. (This "vote-counting" of significant and non-significant results – as a method of summarizing research results – has been superseded by meta-analysis.)

Winne concluded that:

> As a set, the experiments reviewed here no doubt exhibit moderately strong population [generalizability to other kinds of students] and ecological [generalizability to other kinds of settings] validity. . . . Contrastingly, the internal validity ["the extent to which extraneous variables have been controlled by the researcher" (Borg & Gall, 1983, p. 634)] and the integrity of the treatments generally was found to be less adequate. . . . *[T]here is no sturdy conclusion which can be offered here about the relative effectiveness of teachers' use of higher cognitive questions for enhancing student achievement.* [italics added] (pp. 45–46)

Redfield and Rousseau (1981) followed Winne's review with "a meta-analysis [quantitative synthesis] of experimental research on teacher questioning behavior."

Those authors added one experiment to the 18 reviewed by Winne and computed measures of "effect size" (rather than the less informative measures of statistical significance used by Winne; see Kline, 2004). Redfield and Rousseau concluded that their analysis "demonstrates that, regardless of type of study or degree of experimental validity, teachers' predominant use of higher cognitive questions has a positive effect on student achievement" (p. 244).

A second quantitative synthesis of mostly the same literature was performed by Samson, Strykowski, Weinstein, and Walberg (1987). Their conclusion, stemming from modifications in the selection and analysis of the data, was that "higher cognitive questioning strategies have a small positive median effect on learning measures but not as large as has been suggested by the previous meta-analysis [by Redfield and Rousseau (1981)]."

The conclusion from these three reviews of experiments on question-outcome relationships is that the superiority of higher-order questions for promoting achievement is hard to demonstrate. The effect sizes in their favor are unreliable, and teachers who ask higher-order questions are not rewarded by higher student achievement.

Hirsch's Defense of "Knowledge Questions." Hirsch (1987) questioned the value of higher-order questions. He reasoned that so-called lower-order questions that ask students (merely?) for their knowledge (things remembered) have greater educational value than the supporters of higher-order objectives have granted them. He argued that the higher intellectual skills depend on specific knowledge and are therefore specific to particular domains.

In Hirsch's view, critical thinkers and problem solvers in one field, such as mathematics or history, usually fall short in other fields because the thinkers lack the necessary *knowledge*. The chess champion may be a duffer in politics; the brilliant logician may be simple-minded about ethics. The reason for this specificity of critical thinking ability is that enormous amounts of time are necessary for one to acquire the knowledge needed for expertise in any complex domain. So it becomes unrealistic to hope for any great accomplishments as a result of general, or non-domain-specific, training in critical thinking and problem solving. Training that produces good thinkers across many different domains is not yet validated.

Accordingly, the importance of knowledge, defined as ability to remember (recall or recognize) something heard, read, or otherwise experienced, should not be disparaged. Hirsch (1987), acting on his appreciation of knowledge, formulated the concept of "cultural literacy: specific knowledge needed by every member of a society for participating fully in its affairs." The following sample of items from Hirsch's 62-page list of about 2,600 items illustrates what, in Hirsch's conception of cultural literacy, adult Americans should know about:

Archimedes, biofeedback, cabinet (government), composite materials, *Death of a Salesman,* diffraction, Emancipation Proclamation, The grass is always greener on the other side, Grimm Brothers, Joshua, *memento mori* [a reminder of death, such as a skull], paragraph, relative humidity, Stalinism, V-E Day.

All this is not to de-emphasize the importance of the "higher" cognitive objectives. Rather, it recognizes the inordinate difficulty of seeking to achieve such objectives without insuring that the student has the prior or concurrent body of knowledge on which those higher mental processes can operate.

Wait-Time in Asking Questions. Another aspect of soliciting is *wait-time*, defined by Rowe (1974) as being of two kinds:

> Wait-time 1—the number of seconds a teacher waits *after asking a question* before calling on a student to answer it, and Wait-time 2—the number of seconds after the student's response *before* the teacher reacts in some way.

Rowe reported that the average teacher's two wait-times tended to be exceedingly short, averaging 1 second. She found that increasing wait-times by a few seconds tended to (a) increase students' response length, (b) elicit more unsolicited appropriate responses, (c) improve student confidence, (d) raise the frequency of speculative responses, (e) enhance the thoughtfulness of responses, and (f) increase the frequency of student-to-student data comparisons, evidence-inference statements, student questions, and responses from relatively slow students.

Giaconia (1988) found that, in her sample of nine fifth- and sixth-grade teachers, Rowe's conclusions about the duration of wait-time were not entirely accurate. Giaconia's technology for measuring wait-time yielded, for the nine teachers she recorded, averages from 1.45 to 3.45 second for wait-time 1 and 0.63 to 2.15 second for wait-time 2. "Most of these wait-times were about twice as large as the 'typical' wait-time 1 value of 1.0 second reported by Rowe (1974)" (Giaconia, 1988, p. 251).

But the important question is not whether these differences in fractions of a second between Rowe's and Giaconia's findings make a difference in student performance and achievement. Rather it is whether *any* extended wait-time 1 or 2 of the kind first studied by Rowe (1974) makes a difference in student behavior or achievement as compared with that of students whose teachers were unaware of the issue of wait-time.

Giaconia's (1988) descriptive, that is, non-interventional, study, reported that "The role of wait-time appeared subordinate to that of question characteristics in predicting these [desirable] response characteristics" (p. 253). The response characteristics were syntactical complexity, length, and cognitive level of students' responses to questions. That is, questions that had higher cognitive levels, were divergent (i.e., had several correct answers) rather than convergent (had only one correct answer), and open-ended rather than closed-ended, had a greater effect on the students' response characteristics than wait-time did.

Example of the phenomenon to be explained: Pond and Newman (1988) found that a wait-time of five seconds increased the number of correct responses to textually implicit material on standardized reading tests but not for textually explicit material. The authors surmised that the effect resulted from the students' awareness of and involvement in the strategy. They also concluded that student awareness of the wait-time strategy, from hearing explanations of it and using it extensively, was advantageous. Similarly, increasing the wait-times 1 and 2 of teachers' soliciting increases the frequency, length, appropriateness, confidence, originality, thoroughness, and speculativeness of students' responses.

The covering law: Cognitive processes, including responding to questions, may be regarded as responses to more or less complex eliciting stimuli, and reaction time – the time interval between the onset of a stimulus and the onset of an overt response – is positively correlated with stimulus complexity.

Sub-Theories of Responding and Reacting

1.3. Responding (RES): "Responding moves bear a reciprocal relationship to soliciting moves and occur only in relation to them. Their pedagogical function is to fulfill the expectation of soliciting moves; thus students' answers to teacher questions are classified as responding moves. (Bellack et al., 1966, p. 4)

1.4. Reacting (REA). "These moves are occasioned by a structuring, soliciting, responding or a prior reacting move, but are not directly elicited by them. Pedagogically, these moves serve to modify (by clarifying, synthesizing, or expanding) and/or to rate (positively or negatively) what has been said previously. Reacting moves differ from responding moves: while a responding move is always directly elicited by a solicitation, preceding moves serve only as the occasion for reactions. Rating by a teacher of a student's response, for example, is designated a reacting move. (Bellack et al., 1966, p. 4)

My treatments of responding and reacting are combined to recognize their close relationship in the practice of teaching. That close relationship makes it awkward to discuss one without turning immediately to the other. Sirotnik (1983) found that

[T]he most frequently occurring single interaction, is one of students responding to the teacher .... This occurs roughly 15 percent and 10 percent of the time at the elementary and secondary levels, respectively .... [L]ess than 5 percent of teachers' time is spent responding to students, which, as will be seen shortly, is less than the percentage of time students are observed initiating interaction with the teacher. (p. 20)

Since students do almost all of the responding, in what way does responding fit into a theory of *teaching*? The answer is that student responses provide teachers with feedback. How teachers perceive and respond to that feedback falls into the category of what a theory of teaching should embrace. And that feedback inevitably shapes how the teacher reacts.The student's response has several characteristics, such as the student's (a) factual correctness versus error, (b) confidence versus hesitancy of response, (c) economy versus superfluity of words and ideas, and (d) originality versus triteness. Each of these provides the teacher with useful information.

Factual Correctness Versus Incorrectness. If the student's response is correct, it tells the teacher one or more of such things as: The student already knew or could do, prior to being taught, what the question asked for. Or the student has just learned what the question asked for. Or the student made a lucky guess. Or the teacher's question was extremely low in difficulty. Or the teacher's prior structuring was comprehended. The teacher may react to the correctness with a positive comment, such as "Good," "Correct," "That's right."

If the student's response is incorrect, the response tells the teacher one or more of such things as: The student had not read the assignment. Or the student was not paying attention. Or the teacher's structuring was not comprehended by the student. Or the student did not understand the textbook.

The covering law: The teacher's reaction to the student's response. How the teacher reacts to the student's response may be understood in terms of *perceptual*

control theory (Powers, 1973; Runkel, 2003). In brief, perceptual control theory holds that persons' behavior is and should be aimed at their control of their *perceptions*. When a person's perceptions are close to a pre-defined reference point set by the person's needs or desires, the perceiver acts so as to maintain the existing situation. When those perceptions are distant from, or unlike, the reference point, the perceiver operates so as to move the perception closer to the reference point, that is, to change the existing situation to something that will be perceived as closer.

Accordingly, suppose a teacher's reference point for what she wants to perceive is a student's correct response. If the student makes an incorrect response, the teacher – trying to optimize her perception – should experience pressure to correct the student's response. She can do so by giving the student an explanation of why the response was incorrect and how the student's subsequent responses to similar questions can be correct. But Sirotnik (1983), reporting on the data from Goodlad's (1984) large-scale observational study, *A Place Called School,* held that

> Providing corrective feedback in combination with additional information designed to help students understand and correct their mistakes *is almost nonexistent.* [italics added] In fact, reinforcement of any kind *is rarely noticed* [italics added] whether in the form of specific task-related acknowledgement and praise or general support and encouragement (p. 20) ....
> In summary, the typical classroom patterns consist of (1) the teacher's explaining or lecturing to the whole class or to a single student, asking direct, factual questions on the subject matter, or monitoring students; and (2) the students' ostensibly listening to the teacher and responding to teacher-initiated interaction. (pp. 20–21)

Thus it is possible to sort these quotations from Sirotnik according to the categories of Bellack et al., (1966). Doing so indicates that the findings of *A Place Called School* (Goodlad, 1984) agree well with the Structuring-Soliciting-Responding-Reacting analysis of classroom teaching constructed by Bellack et al., (1966), and confirmed by Hoetker and Ahlbrand (1969) and Mehan (1979).

We turn now to sub-theories of the content of teaching. This treatment will draw upon the conception of content set forth in Chap. 5. In that chapter, a central concept was *instructional alignment* – the similarity between the content taught and the content assessed.

Sub-Theories of the Content of Teaching

We turn now to a comparable treatment of the content of teaching.

Instructional Alignment

The importance of instructional alignment is both practical and theoretical. The "fairness," or content validity, of an achievement assessment depends on its instructional alignment. The instructional alignment of the content of teaching should be measured against the curriculum, defined as the statement of what should be taught

(a) in a subject matter, (b) during a specific term of teaching, (c) to students of a specified kind. It is the curriculum that should reign over, i.e., serve as the criterion for, the content of both the teaching and the assessment of student achievement. That is, the curriculum should be the criterion by which the validity of the content of teaching and the content of assessment should be judged.

As the criterion, the curriculum should be determined, in the first place, by the nature of the subject matter, e.g., whether it is of the *physical sciences* (chemistry, mathematics, physics), the *biological sciences* (botany, genetics, physiology, zoology), the *humanities* (reading, art, composition, literature, music), or the *social sciences* (anthropology, civics, geography, history, psychology, sociology). These categories determine the kind of knowledge taught and the corresponding objectives at which the teaching should be aimed.

The curriculum should also be affected by the societal and personal values that the teaching should serve. The societal values are those determined by the needs of the society in which the education occurs. The personal values are those of the students and their teacher.

The Teacher's Choice of Content

Although the curriculum prescribes what is to be taught, teachers vary in how closely they follow that prescription. That is, as was noted in Chap. 5, they vary in their *instructional alignment*: the degree to which what they teach is similar to the content of the assessment of their students' achievement Teachers who bring about higher instructional alignment are considered to provide their students with more opportunity to learn the content on which their achievement will be evaluated.

Influences on the Teacher's Instructional Alignment. What factors influence the teachers' instructional alignment? After all, they teach from the same prescribed curriculum and, often, in any given school district, the same textbooks. These factors include (a) the teacher's knowledge of the content and also, as noted in Chap. 5, (b) the teacher's *pedagogical content knowledge* (Shulman, 1986b).

It is likely that the teacher's knowledge and understanding of the content make her steer the content – as expressed in her structurings and solicitations and in the learning materials – in the directions in which she feels strong. Wanting to succeed, the teacher tends to avoid or deemphasize the aspects of the content in which she lacks confidence.

It is not only her knowledge and understanding in the ordinary sense of, say, what is measured by an assessment of achievement over the content. In that sense, primary-grade teachers of arithmetic or reading are, of course, extraordinarily more knowledgeable than their students. Rather it is the teacher's *pedagogical content knowledge* that typically affects a teacher's decisions about what content to select from the curriculum. As Shulman (1986b) put it,

Mere content knowledge is likely to be as useless pedagogically as content-free skill. But to blend properly the two aspects of the teacher's capacities requires that we pay as much attention to the content aspects of teaching as we have recently devoted to the elements of the teaching process (p. 8).

Pedagogical content knowledge comprises knowledge of such things as what of the content students already know and understand; where to begin in the treatment of a new topic; what are pedagogically effective outlines, sequences, and hierarchies of relevant ideas; what are pedagogically effective explanations, analogies, examples, diagrams, and mnemonic devices; and how to anticipate and cope with students' typical questions, difficulties, and errors.

Teachers' Values and Their Treatment of Content. The teacher's personal values also affect her choice and treatment of content. What is meant by values can be understood in terms of a categorization of personal values developed by the German psychologist Eduard Spranger (1928) in his *Types of Men.* Spranger's categories were described in English by Allport, Vernon, and Lindzey (1960) as follows:

(a) *Theoretical,* the discovery of true knowledge and comprehension for their own sake,
(b) *Economic,* usefulness, practicality, the accumulation of wealth,
(c) *Aesthetic,* the creation and appreciation of form, harmony, and beauty,
(d) *Social,* helping other people,
(e) *Political,* acquiring power and leadership, not necessarily in politics, and
(f) *Religous,* achieving mystical or spiritual union with a deity.

The various subject-matter curricula may serve these values in varying degrees. Among these are such affinities as

(a) science, mathematics, and history with *theoretical* values;
(b) economics, business administration, and accounting with *economic* values;
(c) literature, music, and art with *aesthetic* values;
(d) subjects leading to nursing, social work, and medical work with *social* values;
(e) civics, political science, and leadership with *political* values;
(f) ethics and theology with *religious* values.

But the significance of values for education lies not only in the curriculum. Allport et al., (1960) used Spranger's categories to develop a questionnaire for ranking these values in an individual's personality. Such a ranking of a teacher's values, which constitute *presage* variables, affects the subtleties of her processes in teaching the content. When that content is congruent with her values, it is likely that she displays more *enthusiasm* than when teaching content less closely associated with her values. These differences are not lost upon her students. The teacher's enthusiasm has a considerable history of importance in the research on teaching of the last half-century.

Examples of a phenomenon requiring explanation: (a) Rosenshine (1971) brought together five studies in which observers' ratings of teacher enthusiasm correlated .37 to .56 with student achievement. (b) College students who received enthusiastically delivered lectures reported greater intrinsic motivation and experienced higher levels of vitality (Patrick, Hisley, & Kempler, 2000).

*Covering Law 1:*Role theory (e.g., Biddle, 1979; Johnson & Johnson, 1995; Newcomb, 1961) focuses on the teacher's superior status in the classroom, so that the teacher's enthusiasm about what is being taught makes that content have greater value in the perceptions of the students.

Covering Law 2: The teacher behaviors that betoken her enthusiasm – gesturing, varying intonation, moving back and forth, exciting analogies and examples, and humor – arouse student interest by their *contrast* with the typically sober tenor and subject matters of classroom events.

Sub-Theories of Students' Cognitive Capabilities and Motivation

In Chap. 6, we began by noting that cognitive capabilities are customarily classified into two categories: "intelligence" and "prior knowledge." This distinction is based on the degree to which cognitive capability develops (a) as a result of **essentially unplanned** genetic factors and unplanned experience in everyday living in a family and neighborhood (as is true of cognitive capabilities) or (b) as a result of **carefully planned** and executed student learning in the formal setting of a classroom or computer program (as is true of prior knowledge).

Another aspect of the distinction between intelligence and prior achievement is their **alterability.** Intelligence (mental age/chronological age) is hard, well-nigh impossible, to change in any easily applicable way; it is remarkably stable throughout the life span (Bloom, 1964). Prior knowledge, on the other hand, can be changed through education in relatively controllable ways. Students who go to better schools tend to have greater prior knowledge. Improving schools in terms of curriculum, teacher effectiveness, and instructional resources generally increases prior knowledge.

Both kinds of cognitive capability differ from one individual to another. The study of such individual differences began late in the nineteenth century, developed into the still thriving discipline called the "psychology of individual and group differences," and proved to have practical value in such areas as student and employee selection and guidance. Positive correlations were soon found between cognitive capability and both student achievement and employee competence in many important kinds of work. Those correlations are applied in the selection of students and employees.

Teachers quickly become aware of their students' differences in cognitive capability. The differences reveal themselves in the speed and quality with which students can recall, recognize, comprehend, apply, analyze, integrate, and create within the curriculum. Teachers also quickly become aware of the difficulty of teaching in ways that are appropriate for all of their students – students who usually differ significantly in their cognitive capabilities.

A Sub-Theory of Classroom Management

Cronbach (1967) analyzed the problem of adapting instruction to individual differences in the ways shown in concerning Table 8.3, Cronbach wrote:

> Two preliminary remarks are called for. First, these adaptations are by no means mutually exclusive; they can combine in various patterns, and no doubt all of them have a place in the ideal educational system. Second, it is category 3b ["Teach different pupils by different methods"] that is most interesting ... since all the other devices alter administrative arrangements rather than instructional technique. (p. 23)

Psychologists and educators have made many attempts to to develop models of teaching that would cope with the problem of teaching in a classroom full of students differing in cognitive capability. Among these models are computer-assisted instruction, which began in the 1950s and is still being vigorously advocated and promoted. But, as was noted in Chap. 5, it is, according to the evidence brought together by Cuban (2001), "oversold and underused."

Corno (2008) wrote as follows about teachers in our samples ... especially inclined to look out for student characteristics that *might* impede instruction (including qualities such as inattentiveness and unwillingness to participate) – things that could be noticed easily.

> They had mentally documented a repertoire of academic problems presented by (often) hundreds of previous students that had now become telltale signs for adjustment. For example, one elementary teacher was a stickler for organization. She described a male student in her class as "so disorganized that his lack of attention to details actually interfered with her attempts to teach the whole class." The teacher said that this student failed to benefit from his homework reviews because he rarely completed his homework. Notably, she never really "individualized" instruction—she worked with this student *within* the group context, keeping in mind the need to better the student *in order to better the group.* (p. 21–22)

Table 8.3 Patterns of educational adaptation to individual differences

Educational goals	Instructional treatment	Possible modifications to meet individual needs
Fixed	Fixed	1a. Alter duration of schooling by sequential selection.
		1b. Train to criterion on any skill or topic, hence alter duration of instruction.
Options	Fixed within an option	2a. Determine for each student his perspective adult role and provide a curriculum preparing for that role.
Fixed within a course or program	Alternatives provided	3a. Provide remedial adjuncts to fixed "main track" instruction
		3b. Teach different pupils by different methods

Source: Cronbach (1967, p. 24)

As we noted in Chap. 4, the volume entitled *Models of Teaching* (Joyce, Weil, with Calhoun, 2000) described models for which there is no evidence, based on large-scale surveys, of wide adoption by U.S. teachers. Their Chap. 6, "Personality and Learning Styles: Adapting to Individual Differences," formulates the problem as one of coping with students' differences in "integrative complexity." But they do not deal with the *specifics* of how teachers should act according to the differences among students in "integrative complexity."

The conclusion to which the evidence points is that Cronbach's description of how one teacher handled the problem of individual differences among her students is as good an answer as is now available. This "clinical" approach, sketched on page 107 in Chap. 6, calls upon the teacher to make sometimes rapid judgments and decisions based on her acquisition of experience with the process and content of her teaching, and knowledge of the cognitive capabilities of her students. "Clinical" methods are sometimes contrasted with "statistical" methods, as in making predictions about the achievement of students.

Integrating the Sub-Theories

Until now we have taken teaching apart. Now it is time to put it together. The parts have been (a) the teaching's **process**: the teacher's structuring and soliciting, the students' responding, and the teacher's reacting; (b) the **content** of the teaching: its types of knowledge and the cognitive processes applied to that type of knowledge; and (c) the students' cognitive capability and motivation: above-average, average, and below-average.

Putting it together consists of showing how the various elements of process are related to those of content, and how both must be appropriate to the students' cognitive capability and motivation.

One way to put teaching together, that is, to show this integration, or interrelatedness, is to consider *process, content,* and *cognitive capability* as the three dimensions of a rectangular solid, shown in Fig. 8.2.

Rectangular solid showing three dimensions, each divided into segments (a) Process (structuring, soliciting, and reacting), (b) Content (four types of content and two types of cognitive process: knowledge and understanding), and (c) Student Cognitive Capability and Motivation (below average, average, and above average).

We consider the **process** dimension of the solid to be divided into three teacher-performed segments: structuring, soliciting, and reacting. The **content** dimension of the solid has eight segments: one for each of the eight possible pairings of the four types of knowledge (factual, conceptual, procedural, metacognitive), and two types of cognitive process (knowing and understanding). The **cognitive capability and motivation** dimension is arbitrarily divided into three segments: above-average, average, and below-average.

Within this $3 \times 8 \times 3$ solid, we can visualize 72 cells, each of which represents a particular combination of one segment of the three dimensions. Some of these

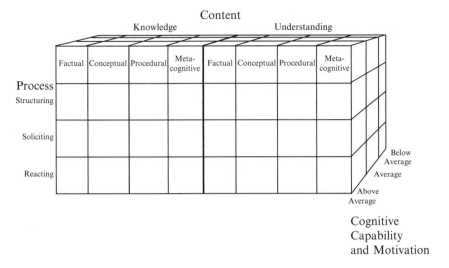

Fig. 8.2 Cognitive capability and motivation.

cells will, in actuality, have many more occurrences of teaching cycles than others. But in theory, each of them is a possible conjunction of a type of process with a type of content for a level of cognitive capability and motivation. Table 8.2 shows examples of teaching cycles consisting of different segments of the three dimensions of teaching.

Example 1. Ms. Eugster may engage in *structuring* a kind of *factual knowledge* that she wants her students of *average cognitive capability* to *remember*. Here the process segment is *structuring*, the content segment is *factual knowledge*, and the cognitive processing segment is *remembering.* The *structuring* consists of informing all her students that the Declaration of Independence was signed on July 4, 1776, and that she wants all her students to *remember* this important date in U.S. history.

Example 2. Ms. Halfpenny may engage in *soliciting* an answer to a question about *conceptual knowledge* from a student of *above-average cognitive capability.* Here the process segment is *soliciting,* the content segment is *conceptual knowledge,* the cognitive process segment is *understanding,* and the cognitive capability segment is *above-average.* The *soliciting* consists of asking an *above-average student* to show his *understanding* by deriving the Pythagorean theorem.

Example 3. Ms. Bollenbacher *reacts* to the erroneous response of a *below-average student* to a question about *procedural knowledge*. Here the *process* segment is reacting, the *content* segment is "how to punctuate a quotation," and *cognitive capability* segment is below-average. The *reacting* consists of saying, "Come on, you can do better. Think of how a quotation looks on the page of a book. Take your time."

Example 4. Ms. Ross *structures* by telling all her students to *apply* the concept of *metacognitive knowledge* to their own studying for a quiz. Here the process segment is *structuring*, the content segment is *metacognitive knowledge*, the *cognitive process* segment question is *apply*, and the *cognitive capability* segment is *average*.

The Culmination: Using the Theory

I intend the theory presented in the foregoing chapters to belong to the behavioral sciences. That is, it is intended to set forth relationships between variables – relationships that cohere as (a) phenomena in teaching that have been or could be observed empirically, and (b) explanations of those phenomena. In the behavioral sciences, such relationships are not as mathematically tight as the relationships in the physical sciences. Rather, as was noted in Chap. 3, they are probabilistic.

The theory should not be regarded as a formula or recipe for good teaching. Rather, as I put it in 1978,

> Teaching is an art — a useful, or practical, art rather than one dedicated to the creation of beauty and the evocation of aesthetic pleasure as ends in themselves. As a practical art, teaching must be recognized as a process that calls for intuition, creativity, improvisation, and expressiveness — a process that leaves room for departures from what is implied by rules, formulas, and algorithms. In teaching, by whatever method it proceeds, there is a need for artistry: in the choice and use of motivational devices, clarifying definitions [and explanations], pace, redundancy, and the like. (p. 15)

Artistic influences operate in many other kinds of work that are governed by bodies of theory. In music, the scores of concertos and symphonies govern performances altogether tightly, but the performers and conductors are permitted, even expected, to modify the performance with artistry—their own subtleties of volume, rhythm, and tempo. The star violinist in a music school follows the notes faithfully, but he does not come close to the artistry of a famous virtuoso, a Jascha Heifetz.

In aeronautics, a graduate student in aeronautical engineering can obey the laws of aerodynamics, just as all engineers must obey those laws. But an American group of engineers designs a four-jet airliner, say, the Boeing 747, that differs in artistic detail from a European group's four-jet Airbus. In poetry, the form of all sonnets must be the same – 14 lines made up of an octave (a stanza of eight lines) and a sestet (the last six lines) – all embodying the statement and resolution of a single theme. Obviously, the restrictions of that form allow abundant room for artistry.

A theory of teaching is analogous to the sonata composer's score, the engineer's laws of aerodynamics, and the poet's rules of some forms of poetry. The teachers who follow the implications of the covering laws used in the present theory of teaching will not lack freedom to adjust their performance to their insights and intuitions. They will be free to depart from the covering laws of our conceptions of the process of teaching, the content of teaching, the cognitive capabilities and motivation of students, and classroom management. As Schön (1983) put it, they can use "reflection-in-action (the 'thinking about what they are doing while they are doing it') that practitioners sometimes bring to situations of uncertainty, uniqueness, and conflict" (p. xi). Scöhn (1987) also wrote of "educating the reflective practitioner." The practitioner's reflections consist of his attempts to merge his knowledge of the theory underlying his practice with whatever artistry can enhance his practice."

The teacher will learn from experience when she should stay close to the implications of the covering laws and when to depart from them. And she will learn from experience whether the structure of the present theory helps her think constructively about her teaching.

References

Adams, R. S. (1971). A sociological approach to classroom research. In I. Westbury & A. A. Bellack (Eds.), *Research into classroom processes: Recent developments and next steps* (pp. 101–117). New York: Teachers College Press.

Alessi, S. M., & Trollip, S. R. (1985). *Computer-based instruction: Methods and development.*Englewood Cliffs, NJ: Prentice-Hall.

Allport, G. W., Vernon, P. E., & Lindzey, G. (1960). *A study of values* (3rd ed.). Boston: Houghton Mifflin.

AMA: American Educational Research Association, Committee on the Criteria of Teacher Effectiveness. (Barr, A. S., Bechdolt, B. V., Coxe, W. W., Gage, N. L., Orleans, J. S., Remmers, H. H., Chairman, & Ryans, D. G.). (1952). Report of the Committee. *Review of Educational Research, 22,* 238–263.

American Institutes for Research. (1999). *An educators' guide to schoolwide reform.* Arlington, VA: Educational Research Service.

Anderson, L. W. (1987). The classroom environment study: Teaching for learning. *Comparative Education Review, 31,* 69–87.

Anderson, L., Evertson, C. M., & Brophy, J. E. (1979). An experimental study of effective teaching in first-grade reading groups. *Elementary School Journal, 79,* 193–223.

Anderson, L. W., Ryan, D. W., & Shapiro, B. J. (Eds.). (1989). The IEA classroom environment study. Oxford: Pergamon.

Anderson, L. W., Krathwohl, D. R., Airasian, P. W., Cruikshank, K. A., Mayer, R. E., Pintrich, P. R., Raths, J., & Wittrock, M. W. (2001). *A taxonomy for learning, teaching, and assessing: A revision of Bloom's taxonomy of educational objectives.* New York: Longman.

Atkinson, R. C., & Shiffrin, R. M. (1968). Human memory: A proposed system and its control processes. In K. W. Spence & J. T. Spence (Eds.), *The psychology of learning and motivation: Advances in research and theory, 2,* (pp. 89–195). New York: Academic Press.

Atkinson, R. L., Atkinson, R. C., Smith, E. E., Bem, D. J., & Nolen-Hoeksema, S. (2000). *Hilgard's introduction to psychology* (13th ed.). New York: Harcourt Brace.

Audi, R. (Ed.). (1995). *The Cambridge dictionary of philosophy.*Cambridge, UK: Cambridge University Press.

Ausubel, D. P. (1963). *The psychology of meaningful verbal learning: An introduction to school learning.* New York: Grune & Stratton.

Ausubel, D. P. (1968). *Educational psychology: A cognitive view.* New York: Holt, Rinehart and Winston.

Ball, S. J. (1995). Intellectuals or technicians? The urgent role of theory in educational studies. *British Journal of Educational Studies, 33,* 255–271.

Banerji, M. (1988). An assessment of the importance of Joseph Mayer Rice in American educational research. 64 pages. ERIC Document Reproduction Service ED294920.

Bar-On, E., & Perlberg, A. (1985). The facet design and smallest space analysis and teachers' instructional behavior. *Studies in Educational Evaluation, 11,* 95–103.

Becker, B. J. (1996). The generalizability of empirical research results. In C. P. Benbow & D. Lubinski (Eds.), *Intellectual talent: Psychometric and other issues* (pp.362–383). Baltimore: Johns Hopkins University Press.

Becker, H. J. (2000). Findings from the teaching, learning, and computing survey: Is Larry Cuban right? *Educational Policy Analysis Archives.* http://epaa.asu.edu/epaa/v8n51

Bellack, A. A. (1976). *Studies in the language of the classroom.* Paper presented at the First Invitational Conference on Teaching. Memorial University of Newfoundland, St. John's, May 25.

Bellack, A. A., Kliebard, H. M., Hyman, R. T., & Smith, F. L. (1966). *The language of the classroom.* New York: Teachers College Press.

Ben-Peretz, M., & Bromme, R. (Eds.). (1990). *The nature of time in schools: Theoretical concepts, practitioner perceptions.* New York: Teachers College Press.

Berger, J., & Zelditch, M. (Eds.). (1993). *Theoretical research programs.* Stanford, CA: Stanford University Press.

Berliner, D. C. (1987). Simple views of effective teaching and a simple theory of classroom instruction. In D. C. Berliner & B. V. Rosenshine (Eds.), *Talks to teachers* 93–110. New York: Random House.

Berliner, D. C. (1989). The place of process-product research in developing the agenda for research on teacher thinking. *Educational Psychologist, 24,* 324–344.

Berliner, D. C. (1990). What's all the fuss about instructional time? In M. Ben-Peretz & R. Bromme (Eds.), *The nature of time in schools: Theoretical concepts, practitioner perceptions* 3–35. New York: Teachers College Press.

Berliner, D. C. (2005). Our impoverished view of educational reform. *Teachers College Record,* published August 2, 2005. Retrieved October 30, 2005, from http://www.tcrecord.org/content. asp?contentid = 12106.

Berliner, D. C., & Biddle, B. J. (1995). *The manufactured crisis.* New York: Addison-Wesley (Republished by Harper Collins).

Beta Blocker Heart Attack Trial Research Group (1982). A randomized trial of propranolol in patients with acute infarction. 1. Mortality results. *Journal of the American Medical Association, 247,* 1707–1714.

Biddle, B. J. (1964). The integration of teacher effectiveness research. In B. J. Biddle & W. J. Ellena (Eds.), *Contemporary research on teacher effectiveness* (pp.1–40). New York: Holt, Rinehart & Winston.

Biddle, B. J. (1979). *Role theory: Expectations, identities, and behaviors.* New York: Academic Press.

Bloom, B. S. (1964). *Stability and change in human characteristics.* New York: Wiley.

Bloom, B. S. (1968). Learning for mastery. *Evaluation comment, 1*(2). Los Angeles: University of California Center for the Study of Evaluation.

Bloom, B. S., Engelhart, M. D., Furst, E. J., Hill, W. H., & Krathwohl, D. R. (1956). *Taxonomy of educational objectives. The classification of educational goals. Handbook 1: Cognitive domain.* New York: Longmans, Green.

Blosser, P. et al. (1973). *A review of research on teacher.* Association for the Education of Teachers in Science, Columbus, OH: ERIC Information Analysis Center for Science, Mathematics, and Environmental Education.

Borg, W. R., & Gall, M. D. (1983). *Educational research: An introduction* (4th ed.). New York: Longman.

Boring, E. G. (1957). *A history of experimental psychology* (2nd ed.). New York: Appleton-Century-Crofts.

Bracey, G. W. (1987a). Measurement-driven instruction: Catchy phrase, dangerous practice. *Phi Delta Kappan, 68,* 683–686.

Bracey, G. W. (1987b). The muddles of measurement-driven instruction. *Phi Delta Kappan, 68,* 688–689.

Brophy, J. E. (2001). *Subject-specific instructional methods and activities.* New York: Elsevier Science.

Brophy, J. E., & Evertson, C. N. (1976). *Learning from teaching: A developmental perspective.* Boston: Allyn and Bacon.

Brophy, J. E., & Good, T. L. (1986). Teacher behavior and student achievement. In M. C. Wittrock (Ed.), *Handbook of research on teaching.* (3rd ed., pp. 328–375). New York: Macmillan.

Broudy, H. (1963). Historic exemplars of teaching method. In N. L. Gage (Ed.), *Handbook of research on teaching* 1–43 Chicago: Rand McNally.

Brown, A. L., & Campione, J. C. (1996). Psychological theory and the design of innovative learning environments: On procedures, burns principles, and systems. In L. Schauble & R. Glaser (Eds.), *Innovations in learning: New environments for education* (pp.289–325). Mahwah, NJ: Erlbaum.

Brown, A. L., & Palincsar, A. S. (1989). Guided cooperative learning and individual knowledge acquisition. In L. B. Resnick (Ed.), *Knowing, learning and instruction: Essays in honor of Robert Glaser.* Hillsdale, NJ: Erlbaum.

Brown, E. Y., Viscoli, C. M., & Horwitz, R. I. (1992). Preventive health strategies and the policy makers' paradox. *Annals of Internal Medicine, 116,* 593–597.

Bruner, J. S. (1966). *Toward a theory of instruction.* Cambridge, MA: Harvard University Press.

Budin, H. (1999). Essay review: The computer enters the classroom. *Teachers College Record, 100,* 656–669.

Burns, R. B. (1984). The process and context of teaching: A conceptual framework. *Evaluation in Education: An International Review Series, 8*(2), 95–112.

Butterfield, H. (1966). *The Whig interpretation of history.* New York: Norton.

Cardinal principles of secondary education (1918). A report of the Commission on the Reorganization of Secondary Education, appointed by the National Education Association. Washington, DC: Government Printing Office.

Carroll, J. B. (1963). A model of school learning. *Teachers College Record, 64,* 723–733.

Carroll, J. B. (1985). The model of school learning: Progress of an idea. In C. W. Fisher & D. C. Berliner (Eds.), *Perspectives on instructional time* (pp. 29–58). New York and London: Longman.

Carroll, J. B. (1989). The Carroll model: A 25-year retrospective and prospective view. *Educational Researcher, 18*(1), 25–31.

Carroll, J. B. (1993). *Human cognitive abilities.* Cambridge, UK: Cambridge University Press.

Case, R. (1985). *Intellectual development: Birth to adulthood.*

Chambers, J. H. (1989). Unacceptable notions of science held by process-product researchers. In R. Page (Ed.), *Proceedings of the forty-fifth annual meeting of the Philosophy of Education Society* (pp. 81–95). Normal, IL: Philosophy of Education Society.

Chambers, J. H. (1992). *Empiricist research on teaching: A philosophical and practical critique of its scientific pretensions.* Boston: Kluwer.

Clandinin, D. J. (1986). *Classroom practice: Teacher images in action.* Philadelphia: Falmer Press.

Clark, C. M., & Peterson, P. L. (1986). Teachers' thought processes. In M. C. Wittrock (Ed.), *Handbook of research on teaching* (3rd ed., pp. 255–296). New York: Macmillan.

Cohen, B. P. (1989). *Developing sociological knowledge: Theory and method* (2nd ed.). Chicago: Nelson-Hall.

Cohen, D. K. (1988). Teaching practice: Plus que ça change. In P. W. Jackson (Ed.), *Contributing to educational change: Perspectives on research and practice* (pp.27–84). Berkeley, CA: McCutchan.

Cohen, S. A. (1987). Instructional alignment: Searching for the magic bullet. *Educational Researcher, 16*(8), 16–20.

Cohen, S. A. (1991). Can fantasies become facts? *Educational Measurement: Issues and Practice, 10*(1), 20–23.

Cohen, S. A. (1995). Instructional alignment. In L. Anderson (Ed.), *International encyclopedia of teaching and teacher education* (2nd ed., pp. 200–204). Tarrytown, NY: Elsevier.

Cook, T. D., Cooper, H., Cordray, D. S., Hartman, H., Hedges, L. V., Light, R. J., Louis, T. A., & Mosteller, F. (1992). *Meta-analysis for explanation: A casebook.* New York: Russell Sage Foundation.

Corno, L. (2008). On teaching adaptively. Unpublished presidential address to Division 15 (Educational Psychology).

Cremin, L. A. (1964). *The transformation of the school: Progressivism in American education, 1876–1957*. New York: Vintage Books, Random House.

Cronbach, L. J. (1955). Processes affecting scores on "understanding of others" and "assumed similarity." *Psychological Bulletin, 52*, 177–193.

Cronbach, L. J. (1967). How can instruction be adapted to individual differences? In R. M. Gagné (Ed.), *Learning and individual differences* (pp. 23–39). Columbus, OH: Charles E. Merrill.

Cronbach, L. J. (1975). Beyond the two disciplines of scientific psychology. *American Psychologist, 30*, 116–127.

Cronbach, L. J. (1986). Social inquiry by and for earthlings. In D. W. Fiske & R. A. Shweder (Eds.), *Metatheory in social science: Pluralisms and subjectivities* (pp. 83–107). Chicago: University of Chicago Press.

Cronbach, L. J., & Snow, R. E. (1977). *Aptitudes and instructional methods: A handbook for research on interactions.*New York: Irvington Publishers, distributed by Halsted Press.

Cronbach, L. J., Gleser, G. C., Nanda, H., & Rajaratnam, N. (1972). *The dependability of behavioral measurements: Theory of generalizability for scores and profiles*. New York: Wiley.

Cuban, L. (1982). Persistence of the inevitable: The teacher-centered classroom. *Education and Urban Society, 15*(1), 26–41.

Cuban, L. (1984). How teachers taught: Constancy and change in American classrooms, 1890–1980. New York: Longman.

Cuban, L. (1988). The managerial imperative and the practice of leadership in school. Albany, NY: State University of New York Press.

Cuban, L. (1986). *Teachers and Machines*. New York: Teachers College Press.

Cuban, L. (1992). *How teachers taught: Consistency and change in American classrooms 1890–1980* (2nd ed.). New York: Longman.

Cuban, L. (1993). Computers meet classroom: Classroom wins. *Teachers College Record, 95*(2), 185–210.

Cuban, L. (2001). *Oversold and underused: Computers in the classroom*. Cambridge, MA: Harvard University Press.

Cutchin, G. C. (1999). Relationships between the big five personality factors and performance criteria for in-service high-school teachers. *Dissertation Abstracts International, Section A: The Humanities and Social Sciences, 59* (7-A, 2263D).

Dearing, E., McCartney, K., & Taylor, B. A. (2001). Change in family income-to-needs matters more for children with less. *Child Development, 72*(6), 1779–1793.

Denham, C. & Lieberman, A. (Eds.). (1980). *Time to Learn*. Washington, DC: National Institute of Education.

Detterman, D. K. (Ed.). (1994). *Theories of intelligence.*Norwood, NJ: Ablex.

Dillon, J. T. (1985). Using questions to foil discussion. *Teaching and Teacher Education, 1*, 109–121.

Dillon, J. T. (1988). The remedial status of student questioning. *Journal of Curriculum Studies, 20*, 197–210.

Dillon, J. T. (1990). *The practice of questioning*. London: Routledge.

Domas, S. J., & Tiedeman, D. V. (1950). Teacher competence: An annotated bibliography. *Journal of Experimental Education, 19*, 101–218.

Doyle, W. (1977). Paradigms for research on teacher effectiveness. In L. S. Shulman (Ed.), *Review of research in education* (pp. 163–198). Itasca, IL: F. E. Peacock.

Doyle, W. (1995).

Dubin, R. (1969). *Theory building*. New York: Free Press.

Dubois, R. W., & Brown, R. H. (1988). Assessing clinical decision making: Is the ideal system feasible? *Inquiry, 28*, 59–84.

Dunkin, M. J., & Biddle, B. J. (1974). *The study of teaching*. New York: Holt, Rinehart & Winston.

Ehrenberg, R. G., & Brewer, D. J. (1995). Did teachers' verbal ability and race matter in the 1960s? Coleman revisited. *Economics of Education Review, 14*(1), 1–21.

Einstein, A. (1951). Autobiography. In P. Schilpp (Ed.), *Albert Einstein: Philosopher-Scientist*, (pp. 683–684). New York: Harper & Row.

Eisner, E. W. (2004). "Artistry in teaching", *Cultural Commons*, http://www.culturalcommons. org/eisner.htm/2span. Accessed: February 11, 2005.

Emmer, E. T., Sanford, J. P., Clements, B. S., & Martin, J. (1982). *Improving classroom management and organization in junior high schools.* (ERIC Document Reproduction Service No. ED261053)

Emmer, E. T., Evertson, C. M., Sanford, J. P., Clements, B. S., & Worsham, M. E. (1984). *Classroom management for secondary teachers.* Englewood Cliffs, NJ: Prentice-Hall.

Engelmann, S. (1980). *Direct instruction.* Englewood Cliffs, NJ: Educational Technology Publication.

English, H. B., & English, A. C. (1958). *A comprehensive dictionary of psychological and psychoanalytical terms: A guide to usage.* New York: David McKay.

Ennis, R. H. (1969). *Logic in teaching.* Englewood Cliffs, NJ: Prentice-Hall.

Evers, W. M. (1998). *What's gone wrong in America's classrooms.* Stanford, CA: Hoover Institution Press.

Evertson, C. M., & Weinstein (2006). *Looking into learning-centered classrooms: Implications for classroom management.* Washington, DC: NEA.

Eysenck, H. J. (1995). Meta-analysis squared—Does it still make sense? *American Psychologist, 50,* 110–111.

Fabes, R. A., Martin, C. L., Harish, L. D., & Updegraff, K. A. (2000). Criteria for evaluating the significance of developmental research in the twenty-first century: Force and counterforce. *Child Development, 71,* 212–221.

Faust, D., & Meehl, P.E. (1992). Using scientific methods to resolve enduring questions within the history and philosophy of science: Some illustrations. *Behavior Therapy, 23,* 195–211.

Festinger, L. (1957). *A theory of cognitive dissonance.* Stanford, CA: Stanford University Press.

Feyerabend, P. (1963). How to be a good empiricist—A plea for tolerance in matters epistemological. In P. H. Nidditch (Ed.), *Philosophy of science* (pp. 12–39). Oxford, UK: Oxford University Press.

Feyerabend, P. (1993). *Against method* (3rd ed.). New York: Verso.

Finn, C. E. (2000). A statement for the B. Fordham Foundation. In *Two paths to quality teaching: Implications for policy makers.* Cheyenne, WY: Education Commission of the States.

Fisch, S. M., & Truglio, R. T. (Eds.). (2001). *"G" is for growing: Thirty years of research on children and Sesame Street.* Mahwah, NJ: Erlbaum.

Fisher, C. W., & Berliner, D. C. (1985). *Perspectives on instructional time.* New York: Longman.

Flanders, N. (1970). *Analyzing teacher behavior.* Reading, MA: Addison-Wesley.

Fosnot, C. T. (Ed.). (1996). *Constructivism: theory, perspectives, and practice.* New York: Teachers College Press.

Gage, N. L. (1960). Metatechnique in educational research. Proceedings, Research Résumé, Twelfth Annual Conference on Educational Research. (pp. 1–12). Burlingame, CA: California Teachers Association.

Gage, N. L. (1963). Paradigms for research on teaching. In N. L. Gage (Ed.), *Handbook of research on teaching* (pp. 94–141). Chicago: Rand-McNally.

Gage, N. L. (1969). Teaching methods. In R. L. Ebel (Ed.), *Encyclopedia of educational research* 4th ed., pp. 1446–1458. London: Macmillan.

Gage, N. L. (1978). *The scientific basis of the art of teaching.* New York: Teachers College Press.

Gage, N. L. (1979). The generality of dimensions of teaching. In P. L. Peterson & H. J. Walberg (Eds.), *Research on teaching: Concepts, findings, and implications.* (pp. 264–288). Berkeley, CA: McCutchan Publishing Corp.

Gage, N. L. (1994a). The scientific status of research on teaching. *Educational Theory, 44,* 371–383.

Gage, N. L. (1994b). The scientific status of the behavioral sciences. An essay-review of Empiricist research on teaching: A philosophical and practical critique of its scientific pretensions by J. H. Chambers. *Teaching and Teacher Education, 10*, 556–577.

Gage, N. L. (1996). Confronting counsels of despair for the behavioral sciences. *Educational Researcher, 25*(1), 5–15, 22.

Gage, N. L., & Needels, M. C. (1989). Process-product research on teaching: A review of criticisms. *Elementary School Journal, 89*(3), 253–300.

Gage, N. L., & Unruh W. (1967). Theoretical formulations for research on teaching. *Review of Educational Research, 37*, 358–370.

Gage, N. L., & Viehover, K. (Eds.). (1975). *Reports of the National Conference on Studies in Teaching, National Institute of Education*. Washington, DC: U.S. Department of Health, Education, and Welfare.

Gagné, R. M. (Ed.). (1985).

Gagné, R. M., & Briggs, L. J. (1979). *Principles of instructional design* (2nd ed.). New York: Holt, Rinehart & Winston.

Gagné, E. D., Yekovich, C. W., & Yekovich, F. R. (1993). *Cognitive psychology of school learning* (2nd ed.). New York: Longman.

Gardner, H. (1983). *Frames of mind: The theory of multiple intelligences*. New York: Basic Books.

Garrison, J. W., & Macmillan. C. J. B. (1984). A philosophical critique of process-product research on teaching. *Educational Theory, 34*, 255–274.

Garrison, J. W., & Macmillan, C. J. B. (1987). Teaching research to teaching practice: A plea for theory. *Journal of Research and Development in Education, 20*(4), 38–43.

Gergen, K. (1973). Social psychology as history. *Journal of Personality and Social Psychology, 26*, 309–320.

Gess-Newsome, J., & Lederman, N. G. (Eds.). (1999). *Examining pedagogical content knowledge: The construct and its implications for science education*. Boston: Kluwer Academic Publishers.

Getzels, J. W., & Jackson, P. W. (1963). The teacher's personality and characteristics. In N. L. Gage (Ed.), *Handbook of research on teaching* (pp. 506–582). Chicago: Rand McNally.

Gewirth, A. (1991). Foreword. In Deryck Beylefeld (Ed.), *The dialectical necessity of morality*. Chicago: University of Chicago Press.

Giaconia, R. M. (1987). *Teacher questioning and wait-time*. PhD. thesis, Stanford University, 1988.

Giaconia, R. M., & Hedges, L. V. (1982). Identifying features of effective open education. *Review of Educational Research, 52*, 579–602.

Glaser, B. G., & Strauss, A. (1967). *The discovery of grounded theory*. Chicago: Aldine.

Glymour, C. (1983). Social science and social physics. *Behavioral Science, 28*, 126–134.

Goldstein, R. E., Andrews, M., Hall, J., & Moss, A. J. (1992). Reduction in long-term cardiac deaths with aspirin after a cardiac event. *Circulation, 86*(Suppl.), 535 [Abstract].

Goodlad, J. M. (1984). *A place called school: Prospects for the future*. New York: McGraw-Hill.

Gore, J. M. (1997). On the use of empirical research for the development of a theory of pedagogy. *Cambridge Journal of Education, 27*(2), 211–221.

Graber, K. C. (2002). Research on teaching in physical education. In V. Richardson (Ed.), *Handbook of research on teaching* (4th ed., pp. 491–519). Washington, DC: American Educational Research Association.

Greer, M. A., Hudson, L. M., & Wiersma, W. (1999). *The Constructivist Teaching Inventory: A new instrument for addressing constructivist teaching practices in the elementary grades*. Paper presented at the meeting of the American Educational Research Association, Montreal, QC, Canada.

Grice, H. P. (1975). Logic and conversation. In P. Cole & J. L. Morgan (Eds.), *Syntax and semantics* Speech acts: *Vol. 3*. (pp. 41–58). New York: Academic Press.

Guttman, L. (1957). A structural theory for intergroup beliefs and action. *American Sociological Review* 24:318–28.

Hanson, N. (1958). *Patterns of discovery: An inquiry into the conceptual foundations of science*. Cambridge, UK: Cambridge University Press.

Harmon-Jones, E., & Mills, J. (Eds.). (1999). *Cognitive dissonance: Progress on a pivotal theory in social psychology*. Washington, DC: American Psychological Association.

Harris, J. R. (1998). *The nurture assumption*. New York: The Free Press.

Hativa, N. (2000). *Teaching for effective learning in higher education*. Boston: Kluwer Academic Publishers.

Hedges, L. V. (1987). How hard is hard science? How hard is soft science? The empirical cumulativeness of research. *American Psychologist, 42*, 443–455.

Hedges, L. V., & Olkin, I. (1985). *Statistical methods for meta-analysis*. New York: Academic Press.

Hellman, H. (1998). *Great feuds in science: Ten of the liveliest disputes ever*. New York: Wiley.

Hempel, C. G. (1965). *Aspects of scientific explanation*. New York: The Free Press.

Hempel, C. G., & Oppenheim, P. (1948). Studies in the logic of explanation. *Philosophy of Science, 15*, 567–579.

Hirsch, E. D., Jr. (1987). *Cultural literacy: What every American needs to know*. Boston: Houghton Mifflin.

Hirsch, E. D., Jr. (1996). *The schools we need and why we don't have them*. New York: Doubleday.

Hirsch, E. D., Jr. (2002). Classroom research and cargo cults. *Policy Review*, No. 115, 51–69.

Hmelo-Silver,C. E., Duncan, R. G., & Chinn, C. A. (2007). Scaffolding and achievement in problem-based and inquiry learning. *Educational Psychologist, 42*(2), 91–97.

Hoetker, J., & Ahlbrand, W. P., Jr. (1969). The persistence of the recitation. *American Educational Research Journal, 6*, 145–167.

Huh, K. C. (1985). The role of teacher logic and clarity in student achievement. *Dissertation Abstracts International, 471*(2), 472.

Hume, D. (1758). *An enquiry concerning human understanding*. Oxford, UK: Clarendon Press.

Husén, T. (1967). *International study of achievement in mathematics: A comparison of twelve countries (Vol. 2)*. New York: Wiley.

Jackson, P. W. (1968). *Life in classrooms*. New York: Holt, Rinehart, and Winston.

Jackson, P. W. (Ed.). (1992). *Handbook of research on curriculum*. New York: Macmillan.

Johnson, M., & Brooks, H. (1979). Conceptualizing classroom management. In D. Duke (Ed.), *Classroom management (The 78th Yearbook of the National Society for the Study of Education, Part)* 2 (pp. 1–41). Chicago: University of Chicago Press.

Johnson, D. W., & Johnson, R. T. (1995). Social psychological theories of teaching. In L. Anderson (Ed.), *International encyclopedia of teaching and teacher education* (2nd ed., pp. 112–117). Oxford: Elsevier.

Joyce, B., Weil, M. (with Calhoun, E.) (2000). *Models of teaching* (6th ed.). Boston: Allyn and Bacon.

Karl Popper (1963). *Conjectures and Refutations: The Growth of Scientific Knowledge*. Routledge, London, 1963.

Kanterovich, A., & Néeman, Y. (1989). Serendipity as a source of evolutionary progress in science. *Studies in History and Philosophy of Science, 20*, 505–529.

Keller, F. (1968). Goodbye, teacher. *Journal of Applied Behavior Analysis, 1*(1), 79–89.

Kerr, D. (1983). Teacher competence and teacher education in the United States. In L. S. Shulman & G. Sykes (Eds.), *Handbook of teaching and policy*. (pp. 126–149). New York: Longman.

Kirschner, P. A., Sweller, J., & Clark, R. E. (2006). Why minimal guidance during instruction does not work: An analysis of the failure of constructivist, discovery, problem-based, experiential, and inquiry-based teaching. *Educational Psychologist, 41*(2), 75–86.

Klahr, D., & Nigam, M. (2004). The equivalence of learning paths in early science instruction: Effects of direct instruction and discovery learning. *Psychological Science, 15*, 661–667.

Kline, R. B. (2004). *Beyond significance testing*. Washington, DC: American Psychological Association.

Koch, S. (1959–1963). *Psychology: A study of a science* (Vols. 1–6). New York: McGraw-Hill.

Koczor, M. L. (1984). Effects of varying degrees of instructional alignment in posttreatment tests on mastery learning tasks of fourth-grade children. (Doctoral dissertation, University of San Francisco, 1984). *Dissertation Abstracts International–A, 46/05*, 1179.

Kolata, G. (1984, January 27). Lowered cholesterol decreases heart disease. *Science, 223*, 581–382.

Kramarski, B., & Mevarech, Z. R. (2003a). The effects of metacognitive instruction on solving mathematical authentic tasks. *Educational Studies in Mathematics, 49*(2), 225–250.

Kramarski, B., & Mevarech, Z. R. (2003b). Enhancing mathematical reasoning in classroom: The effects of cooperative learning and the metacognitive training. *American Educational Research Journal, 40*(1), 281–310.

Krathwohl, D. R., Bloom, B. S., & Masia, B. B. (1964). *Taxonomy of educational objectives: Handbook II: The affective domain.* New York: David McKay.

Kuhn, T. (1962). *The structure of scientific revolutions* (2nd ed., enlarged). Chicago: University of Chicago Press.

Kuhn, T. (1970). *The structure of scientific revolutions* (2nd ed., enlarged). Chicago: University of Chicago Press.

Kulik, J. A. (1981). *Integrating findings from different levels of instruction.* (ERIC Document Reproduction Service No. ED 208040).

Labaree, D. F. (2004). *The trouble with Ed schools* .New Haven, CN:Yale University Press.

Lakatos, I. (Ed.). (1968). *The problem of inductive logic: Proceedings of the International Colloquium in the Philosophy of Science, London, 1965, Vol. 2.* Amsterdam: North-Holland Publishing Company.

Lamm, Z. (1976). *Conflicting theories of instruction: Conceptual dimensions.* Berkeley, CA: McCutchan.

Lazear, E. P., et al. (1992). Prices and wages in transition economies. Stanford, CA: Hoover Institution, Stanford University.

Lemke, M., et al. (2003). International Outcomes of Learning in Mathematics Literacy and Problem Solving: PISA 2003 Results from the U.S. Perspective. *Education Statistics Quarterly, 6*(4).

Levine, M. (1975). *A cognitive theory of learning: Research on hypothesis testing.* Hillsdale, NJ: Halsted Press.

Lewin, K. (1931). The conflict between Aristotelian and Galilean modes of thought in contemporary psychology. *Journal of General Psychology, 5*(2), 143–177.

Lilienthal A. M., Pedersen, E., & Dowd, J. E. (1967). *Cancer epidemiology: Methods of study.* Baltimore: Johns Hopkins University Press.

Lipid Research Clinics Program. (1984). The Lipid Research Clinics primary prevention trial results. 1. Reduction in incidence of coronary heart disease. *Journal of the American Medical Association, 251,* 351–364.

Lipsey, M. W., & Wilson, D. B. (1993). The efficacy of psychological, educational, and behavioral treatment: Confirmation from meta-analysis. *American Psychologist, 48,* 1181–1209.

Lipsey, M. W., & Wilson, D. B. (1995). The role of method in treatment effect estimates: Evidence from meta-analyses of psychologically-based interventions. Paper presented at the NIMH Mental Health Services Research Conference, Bethesda, MD.

Losee, J. (1980). *A historical introduction to the philosophy of science* (2nd ed.). Oxford, UK: Oxford University Press.

Lumsdaine, A. A., & Glaser, R. (Eds.). (1960). *Teaching machines and programmed learning: A source book.* Washington, DC: National Education Association.

MacIntyre, A. (1985). After virtue: A study in moral theory. London: Duckworth.

Macmillan, C. J. B., & Garrison, J. W. (1984). Using the "new philosophy of science" in criticizing current research traditions in education. *Educational Researcher, 13*(10), 15–21.

Macmillan, C. J. B., & Garrison, J. W. (1988). A logical theory of teaching: Erotetics and intentionality. Boston: Kluwer.

Mandeville, G. K. (1989). Descriptions of lessons. In L. Anderson, . (Eds.), *The IEA classroom environment study* (pp. 127–145). Elmsford, NY: Pergamon Press.

Marland, P. W. (1995). Implicit theories of teaching. In L. Anderson (Ed.), *International encyclopedia of teaching and teacher education* (2nd ed., pp.131–136). Oxford, UK: Elsevier.

Marrow, A. J. (1969). *The practical theorist: The life and work of Kurt Lewin.* New York: Basic Books.

Martin, J., & Sugarman, J. (1993). Beyond methodolatry: Two conceptions of relations between theory and research in research on teaching. *Educational Researcher, 22*(8), 17–24.

Marx, R. W., & Winne, P. H. (1987). The best tool teachers have—Their students' thinking. In D. C. Berliner & B. V. Rosenshine (Eds.), Talks to teachers (pp. 267–304). New York: Random House.

Masterman, M. (1970). The nature of a paradigm. In I. Lakatos & A. Musgrave (Eds.), Criticism and the growth of knowledge (pp. 59–89). New York: Cambridge University Press.

McConnell, J. W., & Bowers, N. D. (1979). A comparison of high-inference and low-inference measures of teacher behaviors as predictors of pupil attitudes and achievements. (ERIC Document Reproduction Service No. ED171780).

McLaughlin, M. W., & Talbert, J. E. (2001). Professional communities and the work of high school teaching. Chicago: University of Chicago Press.

McLaughlin, T. F., & Williams, R. L. (1988). The token economy. In J.C. Witt, S. N. Elliott, & F. M. Gresham (Eds.), Handbook of behavior therapy in education (pp. 469–487). New York: Plenum Press.

Medawar, P. B. (1984). The limits of science. New York: Harper and Row.

Medley, D. M. (1977). Teacher competence and teacher effectiveness: A review of process-product research. Washington, DC: American Association of Colleges for Teacher Education.

Medley, D. M., & Mitzel, H. E. (1963). Measuring classroom behavior by systematic observation. In N. L. Gage (Ed.), Handbook of research on teaching (pp. 247–328). Chicago: Rand McNally.

Mehan, H. (1979). Learning lessons: The social organization of the classroom. Cambridge, MA: Harvard University Press.

Merrow, J. (2004). Can D. C.'s search make the grade? The Washington Post, August 8, Page B-01.

Merton, R. K. (1955). A paradigm for the study of the sociology of knowledge. In Lazarsfeld, P. F. & Rosenberg, M. (Eds.), The language of social research. (pp. 498–510). Glencoe, IL: Free Press.

Mitchell, W. T. (Ed.). (1985). Against theory: Literary studies and the new paradigm. Chicago: University of Chicago Press.

Mitzel, H. E. (1957). A behavioral approach to the assessment of teacher effectiveness. New York: Division of Teacher Education, College of the City of New York, (Mimeographed).

Mitzel, H. E. (1960). Teacher effectiveness. In C. W.Harris (Ed.), Encyclopedia of educational research (3rd ed., pp. 1481–1486). New York: Macmillan.

Monk, D. H. (1994). Subject area preparation of secondary mathematics and science teachers and student achievement. Economics of Education Review, 13(2), 125–145.

Morsh, J. E., & Wilder, E. W. (1954). Identifying the effective instructor: A review of the quantitative studies: 1900–1953, Research Bulletin AFPATRC-TR-54-44, Lackland Air Force Base, TX: Air Force Personnel and Training Research Center.

Nagel, E. (1977). Review of Against Method by Paul Feyerabend. American Political Science Review, 71, 12–39.

Nagel, E. (1979). The structure of science: Problems in the logic of scientific explanation (2nd ed.). Indianapolis, IN: Hackett.

National Center for Education Statistics. (2000). Digest of education statistics. Washington, DC: Office of Educational Research and Improvement, U. S. Department of Education.

Needels, M. C. (1984, September). The role of logic in the teacher's facilitation of student achievement. Dissertation Abstracts International-A 4503, 791.

Needels, M. C. (1988). A new design for process-product research on the quality of discourse in teaching. American Educational Research Journal, 25(4), 503–526.

Needels, M. C., & Gage, N. L. (1991). Essence and accident in research on teaching. In H. C. Waxman & H. J. Walberg (Eds.), Effective teaching: Current research 3–31. Berkeley, CA: McCutchan.

Neisser, U. (1998). The rising curve: Long-term gains in IQ and related measures. Washington, DC: American Psychological Association.

Newby, T. J. (1991). Classroom motivation: Strategies of first-year teachers. Journal of Educational Psychology, 83(2), 195–200.

Newcomb, T. M. (1950). Social psychology.New York: Dryden Press.

Newcomb, T. M. (1961). The acquaintance process. New York: Holt, Rinehart and Winston.

Nichols, S. L., & Berliner, D. C. (2007). *Collateral damage: How high-stakes testing corrupts America's schools*. Cambridge, MA: Harvard Education Press.

Nuthall, G. (2005). The cultural myths and realities of classroom teaching and learning. *Teachers College Record, 107*(5), 895–934.

Nuthall, G. (2007). *The hidden lives of learners*. Wellington, New Zealand: New Zealand Council for Educational Research.

Nuthall, G. A., & Alton-Lee, A. G. (1998). *Understanding learning in the classroom: understanding learning and teaching project 3*. Report to the Ministry of Education. Wellington: Ministry of Education.

Nuthall, G., & Alton-Lee, A. (1993). Predicting learning from student experience of teaching: A theory of student knowledge construction in classrooms. *American Educational Research Journal, 30*(4), 799–840.

Nuthall, G., & Snook, I. A. (1973). Models in educational research. In R. M. W. Travers's *Second handbook of research on teaching; A project of the American Educational Research*

Othanel Smith, B., Meux, M., Coombs, J., Nuthall, G.A., & Precians, R. (1967). *A study of the Strategies of teaching*. Urbana: Bureau of Educational Research, University of Illinois.

Patrick, B. C., Hisley, J., & Kempler, T. (2000). 'What's everybody so excited about?': The effects of teacher enthusiasm on student intrinsic motivation and vitality. *Journal of Experimental Education*.

Pelgrum, W. J. (1989). *Educational assessment: Monitoring, evaluation and the curriculum*. Unpublished doctoral dissertation. Enschede, The Netherlands: University of Twente.

Pelgrum, W. J., Eggen, T. H. J. M., & Plomp, TJ. (1986). *The implemented and attained mathematics curriculum: A comparison of eighteen countries*. Enschede, The Netherlands: T.H. (prepared for the National Council for Educational Statistics, Washington, DC).

Phillips, D. C. (1985). *Perspectives on learning*. New York: Teachers College, Columbia University.

Phillips, D. C., & Burbules, N. C. (2000). *Postpositivism and educational research*.Lanham, MD: Rowman & Littlefield Publishers.

Piaget, J., & Inhelder, B. (1973). *Memory and intelligence*. New York: Basic Books.

Pitt, Joseph C. (Ed.). (1988). *Theories of explanation*. Oxford, UK: Oxford University Press.

Pond, M. R., & Newman, I. (1988). *Differential effects of wait-time on textually explicit and implicit responding: Interactional Explanation*. (ERIC 293864.)

Popham, W. (1993). Measurement-driven instruction as a "quick-fix" reform strategy. *Measurement and Evaluation in Counseling and Development, 26*(9), 31–34.

Popham, W. J., Cruse, K. L., Rankin, S. C., Sandifer, P. D., & Williams, P. L. (1985). Measurement-driven instruction: It's on the road. *Phi Delta Kappan, 66*, 628–634.

Popkewitz (1984). Paradigm and ideology in educational research: The social functions of the intellectual. New York: Falmer Press.

Popper, K. (1963). *Conjectures and refutations: The growth of scientific knowledge*. Routledge, London.

Popper, K. (1965). *Conjectures and refutations: The growth of scientific knowledge* (2nd ed.). Routledge & Kegan Paul.

Porter, A. C. (2002). Measuring the content of instruction: Uses in research and practice. *Educational Researcher, 31*(7), 3–14.

Powers, W. T. (1973). *Behavior: The control of perception*. New York: Aldine.

Rajagopalan, K. (1998). On the theoretical trappings of the thesis of anti-theory; or, why the idea of theory may not, after all, be all that bad: A reply to Gary Thomas. *Harvard Educational Review, 68*, 335–352.

Random House Webster's College Dictionary. (1991). New York: Random House.

Redfield, D. L., & Rousseau, E. W. (1981). A meta-analysis of experimental research on teacher questioning behavior. *Review of Educational Research, 51*, 237–245.

Reigeluth, C. M. (1999). *Instructional-design theories and models: A new paradigm of instructional theory*,Vol. 2. Mahwah, NJ: Erlbaum.

Rice, J. M. (1897). The futility of the spelling grind. Reprinted in J. M. Rice (1913). *Scientific management in education* (pp. 65–99). New York: Hinds, Noble & Eldredge.

Richardson, V. (1997). *Constructivist teacher education.* Washington, DC: The Falmer Press.

Richardson, V. (Ed.). (2001). *Handbook of research on teaching* (4th ed.). Washington, DC: American Educational Research Association.

Ridenour, L. (1950). *Educational research and technological change* . Paper presented at the annual meeting of the American Educational Research Association. Washington, DC: American Educational Research Association.

Rosenshine, B. (1971). *Teaching behaviours and student achievement.* Slough, UK: National Foundation for Educational Research in England and Wales.

Rosenshine, B. V. (1987). Explicit teaching. In D. C. Berliner & B. V. Rosenshine (Eds.), *Talks to teachers* 75–92. New York: Random House.

Rosenshine, B., & Meister, C. (1995). Direct instruction. In L. Anderson (Ed.), *International encyclopedia of teaching and teacher education.* (2nd ed., pp. 143–149). Tarrytown, NY: Elsevier Science.

Rosenthal, R. (1994). Parametric measures of effective size. In H. Cooper & L. V. Hedges (Eds.), *Handbook of research synthesis* (pp. 231–244). New York: Russell Sage Foundation.

Rothenberg, J. (1989). The open classroom reconsidered. *Elementary School Journal, 90,* 69–86.

Rowe, M. B. (1974). Wait-time and rewards as instructional variables: Their influence on language, logic, and fate-control, Part 1. Wait-time. *Journal of Research in Science Teaching, 11,* 81–94.

Rowe, M. B. (1978). Wait, wait, wait. *School Science and Mathematics, 28*(3), 207–216.

Runkel, P. J. (2003). *People as living things: The psychology of perceptual control.* Hayward, CA: Living Control Systems Publishing.

Ryan, D. W., Hildyard, A., & Bourke, S. (1989). Description of classroom teaching and school learning. In L. W. Anderson, D. W. Ryan, & B. J. Shapiro (Eds.), *The IEA classroom environment study* (pp. 71–126). New York: Pergamon Press.

Ryans, D. G. (1960). *Characteristics of teachers.* Washington, DC: American Council on Education.

Samson, G. E., Strykowski, B., Weinstein, T., & Walberg, H. J. (1987).The effects of teacher questioning levels on student achievement: A quantitative synthesis. The Journal of Educational Research, *80,* 290–295.

Schmidt, W. H. (1978). *Measuring the content of instruction.* Research Series No. 35. East Lansing, MI: Institute of Research on Teaching, Michigan State University.

Schmidt, F. L. (1992). What do data really mean? Research findings, meta-analysis, and cumulative knowledge in psychology. *American Psychologist, 47,* 1173–1181.

Schmidt, H. G., Loyens, S. M. M., Van Gog, T., & Paas, F. (2007). Problem-based learning is compatible with human cognitive architecture: Commentary on Kirschner, Sweller, & Clark (2006). *Educational Psychologist, 42*(2), 91–97.

Schoenfeld, A. H. (1998). Toward a theory of teaching-in-context. *Issues in Education, 4,* 1–94.

Schön, D. A. (1983). *The reflective practitioner.* San Francisco: Jossey-Bass.

Schön, D. A. (1987). *Educating the reflective practitioner.* San Francisco: Jossey-Bass.

Scriven, M. (1968). The philosophy of science. In D. L. Sills (Ed.), *International encyclopedia of the social sciences* (Vol. 14, p. 84). New York: Macmillan and Free Press.

Shulman, L. S. (1986a). Paradigms and research programs in the study of teaching: A contemporary perspective. In M. C.Wittrock (Ed.), *Handbook of research on teaching* (3rd ed., pp. 3–36). New York: Macmillan.

Shulman, L. S. (1986b). Those who understand: A conception of teacher knowledge. *American Educator, 10*(1), 9–15, 43–44.

Shulman, L. S. (1987). Knowledge and teaching: Foundations of the new reform. *Harvard Educational Review, 57*(1), 1–22.

Shulman, L. S., et al. (Chair). (1975). *Teaching as clinical information processing (Report of Panel 6, National Conference on Studies in Teaching, National Institute of Education).* Washington, DC: U.S. Department of Health, Education and Welfare.

Sirotnik, K. A. (1983). What you see is what you get—Consistency, persistency, and mediocrity in classrooms. *Harvard Educational Review, 53*, 16–31.

Skinner, B. F. (1938). *The behavior of organisms* (7th printing). New York: Appleton-Century-Crofts.

Skinner, B. F. (1950). Are theories of learning necessary? Psychological Review, 57, 193-216

Skinner, B. F. (1953). *Science and human behavior*. New York: MacMillan.

Skinner, B. F. (1957). Schedules of reinforcement, with C.B. Ferster, 1957. ISBN 0-13-792309-0.

Skinner, B. F. (1964). Skinner on theory. *Science, 145*, 1385, 1387.

Skinner, B. F. (1968). *The technology of teaching*. New York: Appleton-Century-Crofts.

Skinner, B. F. (1974). *About behaviorism*. New York: Vintage Books.

Slavin, R. (1990). *Cooperative learning: Theory, research, and practice*. Englewood Cliffs, NJ: Prentice-Hall.

Smith, B. O. (1961). A concept of teaching. In B. O. Smith & R. Ennis (Eds.) , *Language and concepts in education* (pp. 86–101. Chicago: Rand McNally.

Smith, B. O. (1963). Toward a theory of teaching. In A. A. Bellack (Ed.), *Theory and research in teaching* (pp. 1–10). New York: Bureau of Publications, Teachers College, Columbia University.

Smith, B. O. (1983). Some comments on educational research in the twentieth century. *Elementary School Journal, 83*(4), 488–492.

Smith, B. O., Meux, M. O., Coombs, J., Eierdam, D., &Szoke, R(1962). A study of the logic of teaching. Unpublished study available from the Bureau of Educational Research, College of Education, University of Illinois: Urbana, IL.

Smith, B. O., Meux, M., Coombs, J., Nuthall, G. A., & Precians, R. (1967). *A study of the strategies of teaching*. Urbana: Bureau of Educational Research, University of Illinois, 1967.

Smith, B. O., Meux, M. O., et al. (1967).

Smith, R. B. (1987). *The teacher's book of affective instruction: A competency based approach*. Lanham, MD: University Press of America.

Snow, R. E. (1973). Theory construction for research on teaching In R. M. W. Travers (Ed.), *Second handbook of research on teaching* (2nd ed., pp. 77–112). Chicago: Rand McNally.

Sohn, D. (1995). Meta-analysis as a means of discovery. *American Psychologist, 47*, 108–110.

Spearman, C. (1904). "General intelligence" objectively determined and measured. *American Journal of Psychology, 15*, 201–293.

Spranger, E. (1928). *Types of men: The psychology and ethics of personality*. Halle, Germany: M. Niemeyer.

Stallings, J. A. (1975). Implementation and child effects of teaching practices in Follow Through classrooms. *Monographs of the Society for Research in Child Development, 40*(7–8).

Stallings, J. A., & Kaskowitz, D. (1974) Follow-Through classroom observation evaluation, 1972–1973 (SRI Project NRU–7370) Stanford, CA: Stanford Research Institute.

Stigler, J. W., & Hiebert, J. (1999). *The teaching gap: Best ideas from the world's teachers for improving education in the classroom*. New York: The Free Press.

Stipek, D. (2002). *Motivation to learn: Integrating theory and practice* (4th ed.). Boston: Allyn and Bacon.

Strauss, R. P., & Sawyer, E. A. (1986, February). Some new evidence on teacher and student competencies. *Economics of Education Review Elsevier, 5*(1), 41–48.

Suppes, P. (1966). The uses of computers in education. *Scientific American, 215*(3), 206–216.

Suppes, P. (1974). The place of theory in educational research. *Educational Researcher, 3*(6), 3–10.

Sweller, J. (1999). *Instructional design in technical areas: Australian Education Review No. 3*. Camberwell, Victoria, Australia: Australian Council for Educational Research Press.

Sweller, J., Kirschner, P. A., & Clark, R. E. (2007). Why minimally guided teaching techniques do not work: A reply to commentaries. *Educational Psychologist, 42*(2), 115–121.

Thayer, V. T. (1928). *The passing of the recitation*. Boston: Heath.

Thomas, G. (1997). What's the use of theory? *Harvard Educational Review, 67*(1), 75–104.

Thomas, G. (1999). Hollow theory: A reply to Rajogopalan. *Harvard Educational Review, 69*, 51–66.

Thomas, G. (2007). *Education and theory: Strangers in paradigms*.London: Open University Press.

Travers, R. M. W (Ed.). (1973). *Second handbook of research on teaching: A project of the American Educational Research Association.* Chicago: Rand McNally.

Tukey, J. (1962). The future of data analysis. *Annals of Mathematical Statistics, 33,* 13–14.

Turkheimer, E., Haley, A., Waldron, M., D'Onofrio, B., & Gottesman, I. J. (2003). Socioeconomic status modifies heritability of IQ in young children. *Psychological Science, 14*(6), 623–628.

Tyler, R. W. (1951). The functions of measurement in improving instruction. In E. F. Lindquist (Ed.), *Educational measurement* (pp. 47–67). Washington, DC: American Council of Education.

U. S. Department of Education, National Center for Educational Statistics. (1996). *Pursuing Excellence. A study of US Eighth-grade mathematics and science teaching, learning, curriculum, and achievement in international context.* Boston College, MA: The Third Math TIMSS (Third International Mathematics and Science Study).

VandenBos, G. R. (Ed.). (2007). *APA dictionary of psychology.* Washington, DC: American Psychological Association.

Vygotsky, L. S. (1978). *Mind in society.*Cambridge, MA: Harvard University Press.

Walberg, H. (1986). Synthesis of research on teaching In M. C. Wittrock (Ed.), *Handbook of research on teaching* (3rd ed., pp. 214–229). New York: Macmillan.

Wallen, N. E., & Travers, R. M. W. (1963). Analysis and investigation of teaching methods. In N. L. Gage (Ed.), *Handbook of research on teaching* (pp. 448–505). Chicago: Rand McNally.

Weissler, A. M., Miller, B. L., & Boudoulas, H. (1989). The need for clarification of percent risk reduction data in clinical cardiovascular trial reports. *Journal of the American College of Cardiology, 13,* 764–766.

Wennberg, J. E. (1989). *Outcomes research and the evaluative clinical sciences at Dartmouth.* Hanover, NH: Dartmouth Medical School.

Willer, D., & Willer, J. (1973). *Systematic empiricism: Critique of a pseudoscience.* Englewood Cliffs, NJ: Prentice-Hall.

Winne, P. H. (1979). Experiments relating teachers' use of higher cognitive questions to student achievement. *Review of Educational Research, 49,* 13–50.

Winne, P. H. (1982). Minimizing the black box problem to enhance the validity of theories about instructional effects. *Instructional Science, 11,* 13–28.

Winne, P. H. (1987). Why process-product research cannot explain process-product findings and a proposed remedy: The cognitive-mediational paradigm. *Teaching and Teacher Education, 3,* 333–356.

Winne, P. H. (1995). Information processing theories of teaching.In L. Anderson. (Ed.), *International encyclopedia of teaching and teacher education.* (2nd ed., pp. 107–112). Oxford, UK: Elsevier.

Winne, P. H., & Marx, R. W. (1983). *Students' cognitive processes while learning from teaching.* Vol. 1. (Final Report, NIE-G-79–0098). Burnaby, British Columbia, Canada: Simon Fraser University.

Winne, P. H., & Marx, R. W. (1987).

Wittrock, M. C. (Ed.). (1986a). *Handbook of research on teaching* (3rd ed.). New York: Macmillan.

Wittrock, M. C. (1986b). Students' thought processes. In M. C. Wittrock (Ed.), *Handbook of research on teaching* (3rd ed., pp. 297–314). New York: Macmillan.

Zuckerman, M. (2005). Classroom revolution. *U.S. News & World Report, 139*(13), 68.

Author Index

Subject Index